Life

Life

EXTRAORDINARY ANIMALS, EXTREME BEHAVIOUR

Martha Holmes and Michael Gunton

Rupert Barrington, Adam Chapman,
Patrick Morris, and Ted Oakes

UNIVERSITY OF CALIFORNIA PRESS

Berkeley Los Angeles

Introduction

Right *Chinstrap penguins resting on blue ice off the South Sandwich Islands, Antarctica – an example of a bird that has adapted both its body and its behaviour to meet the challenges of polar conditions.*

Previous page *North American red-winged blackbirds on migration to their breeding sites – a feat made possible by their skills of flight and navigation.*

Page one *A crested black macaque, Sulawesi, Indonesia, as fascinated by the behaviour of the person photographing him as the human primate is by him.*

Page two *A school of migrating cownose rays, Ogasawara, Japan – just one of the 28,000 or so known species of fish.*

Life, both the BBC series and this book that accompanies it, is about behaviour – the extraordinary ends animals and plants go to in order to survive and to pass their genes through a new generation.

Every day, animals and plants face enormous challenges thrown at them by predators, competitors and the environment they live in. For most animals, it is a huge achievement just to survive, to see another dawn. But at some point, they also have to reproduce. To do so, they are likely to face serious competition and must either fight rivals to win a mate or invest in elaborate displays to attract one. *Life* is a collection of some of the most exciting examples we could find to illustrate how different groups of living things endeavour to overcome these universal challenges.

There are, of course, many millions of different creatures living on our planet, and the selection for *Life* represents a minuscule fraction of them. The demands of television have meant that we couldn't represent the entire living world, and so we've chosen a cast of characters that we feel best illustrates the diversity and complexity of life. We had to miss out many creatures that are too small to be seen or whose behaviour is less interesting, and then grouped our selection in the simplest way, into programmes – and chapters – about insects, birds, reptiles and so on. And in some cases, we had to combine groups of animals, such as the marine invertebrates. This selection took years to research and then film, with the help of scientists and field-workers worldwide. And whether it's a capuchin monkey using a stone that it can barely lift to smash open palm nuts, a komodo dragon stalking its prey for weeks, two giant beetles wrestling in the treetops or half a million spider crabs massing together to moult, we have been continually astounded by the extraordinary behaviours we've been lucky enough to film for the series.

The diversity of life on Earth, the only planet known to support life, is astonishing. It is the result of more

Opposite *Horseshoe crabs emerging from the sea to spawn in Delaware Bay. These marine creatures are little changed from ones that existed in the ocean more than 400 million years ago, proving that ancient lifestyles are sometimes the best.*

than 3 billion years of evolution. The millions upon millions of organisms alive today all have a common ancestry in the simplest forms of life – mere carbon compounds swirling around in a cocktail of chemicals. Those first primordial compounds had the ability to replicate themselves, and the basis of life was born.

Throughout the ensuing eons, the complexity of those early organic compounds increased and turned into protein-making compounds, and eventually into the simplest cellular organisms. Teamwork was the next step, when different kinds of simple cells got together to create complex groupings. But throughout this long, drawn-out process, only the cells that were most suited to the environment thrived. The less perfectly adapted cells perished, and the process of natural selection began.

Life forms became increasingly complicated. They developed simple guts, muscle fibres and nervous systems. And when the momentous breakthrough of sexual reproduction occurred, and replication was not simply a matter of cloning but of creating a mix of attributes from different individuals, the potential for diversification vastly increased. So did the potential for the development of new species.

More and more new species evolved, exploited new habitats and niches and continued to adapt. Natural selection meant many species died out along the way, unable to compete or survive the rigours of a changing environment. But those that survived kept developing a little for the better with each generation. And today we have the result – so far: a planet with a breathtaking variety of life.

No one really knows how many species there are today – estimates range widely from 4 million to 100 million – but all have one thing in common: the drive to survive and reproduce. It's this perpetual battle that *Life* is all about.

MARTHA HOLMES AND MICHAEL GUNTON

Left *A high-ranking young Japanese macaque warming up in a hot pool in Japan – a smart way to weather extreme cold.*

Life locations – where the stories were filmed

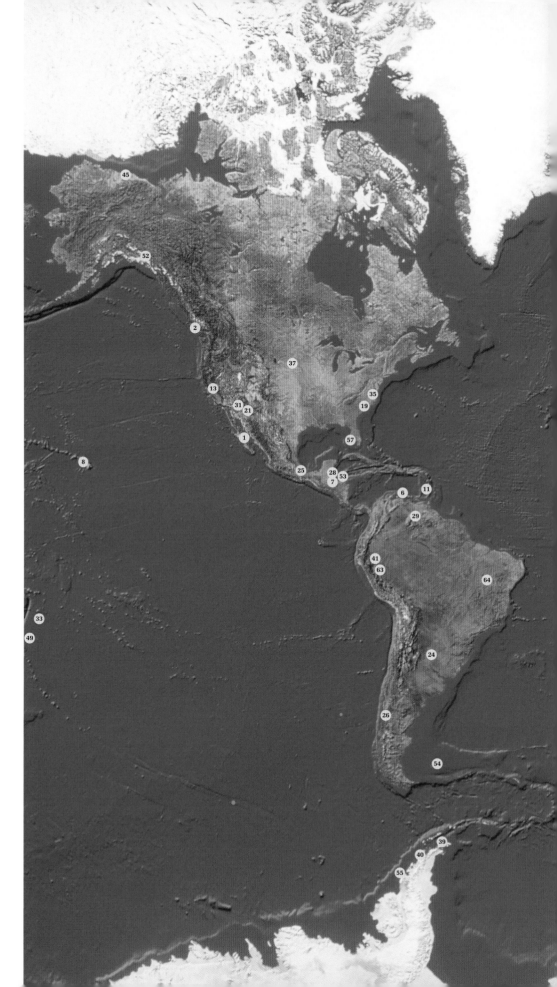

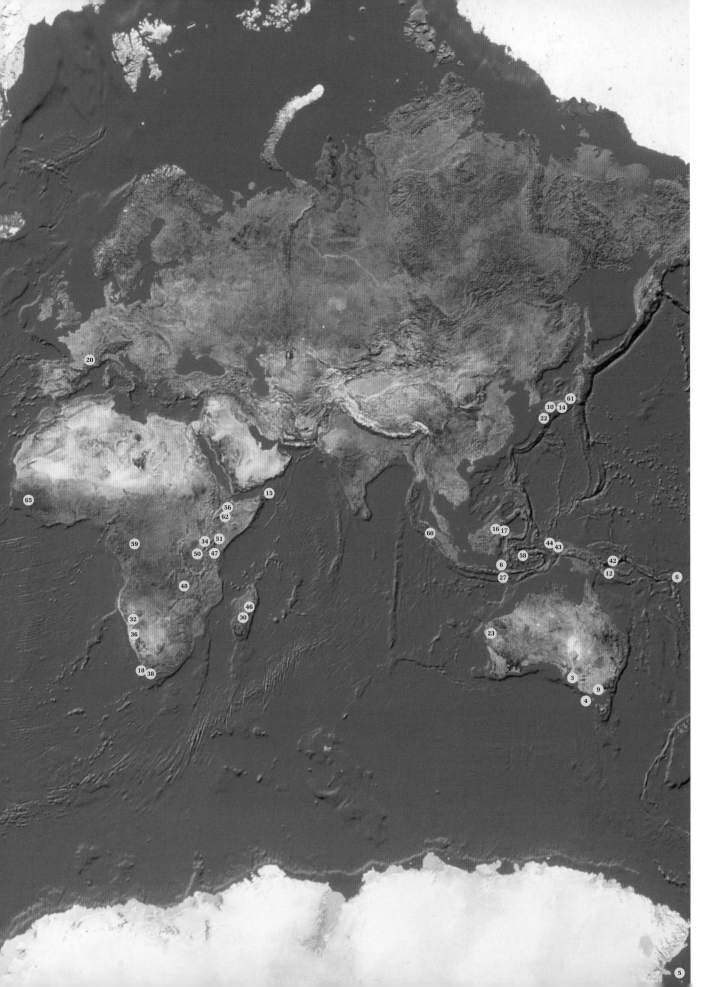

45 | **Polar bears**
Kaktovik, Alaska,
USA

46 | **Aye-ayes**
Antananarivo,
Madagascar

47 | **Rufous sengis**
Rukinga, Kenya

48 | **Straw-coloured fruit bats**
Kasanka National Park,
Zambia

49 | **Humpback whales**
Tonga

50 | **Spotted hyenas**
Serengeti National Park,
Tanzania

51 | **Cheetahs**
Lewa Downs, Isiolo,
Kenya

52 | **Lynx and snowshoe hares**
Haines Junction, Yukon,
Canada

53 | **Bulldog bats**
Belize

54 | **Killer whales and
southern elephant seals**
Sea Lion Island,
Falkland Islands

55 | **Killer whales and
crabeater seals**
Adelaide Island,
Antarctic Peninsula

56 | **Ethiopian wolves**
Bale National Park,
Ethiopia

57 | **Bottlenose dolphins**
Florida Bay, Florida,
USA

58 | **Spectral tarsiers**
Tangkoko Nature
Reserve, Sulawesi,
Indonesia

59 | **Western lowland gorillas**
Ouesso, Democratic
Republic of the Congo

60 | **Orang-utans**
Gunung Leuser National
Park, Sumatra,
Indonesia

61 | **Japanese macaques**
Jigokudani,
Japan

62 | **Hamadryas baboons**
Awash National Park,
Ethiopia

63 | **Red uakari monkeys**
Yavari riverbasin,
Peru

64 | **Brown tufted capuchins**
Barreiras, Brazil

65 | **Chimpanzees**
Bossou,
Guinea

chapter 1

Extraordinary sea creatures

THE EARLIEST FORMS OF LIFE were nurtured in nutrient-rich warm sea water. Over 3 billion years, these marine organisms evolved into the huge variety of plants and animals living today. All living things contain water, and one of the defining features of Earth is the quantity of water that covers its surface. Though much of it is cradled in ocean basins, water still covers approximately 70 per cent of the planet's surface, almost all of it sea water.

Of all the organisms living in the sea, the most diverse and numerous are the invertebrates – literally, those animals without backbones. They include every shape and size of animals, including the sponges (essentially a collection of specialized cells); the cnidarians (which include the sea anemones, corals and jellyfish, all with radial symmetry); the comb jellies, or ctenophores (which have beating cilia – little hairs); all the bilaterally symmetrical worms (including the flatworms, ribbon worms, nematodes and segmented worms); the molluscs (snails, clams, octopuses, indeed, more species than any other group in the sea); the arthropods (the 'insects of the sea', such as barnacles, shrimps, lobsters, crabs); the echinoderms (including sea stars, sea urchins, brittlestars and sea cucumbers) and a collection of other small groups.

One reason the oceans sustain so many animals is that sea water can be occupied far more easily than the air above land. A swimming cuttlefish, for example, is to a large degree supported by the water, whereas an equivalent animal on land would expend a great deal of energy keeping airborne. But even though the available living space for marine organisms is an estimated 250 times greater than for land-dwellers, most marine life is concentrated in the upper 200m (655 feet) where sunlight penetrates. But within this relatively shallow zone, life is not evenly distributed – most marine animals live near land, on or above the continental shelves.

Sunlight and substrate determine the location of the richest communities. Marine plants need sunlight to grow, and a hard, rocky substrate offers them a place to anchor. Around them, complex ecosystems can develop, in tropical, temperate and, to a certain extent, polar seas. Nevertheless, for all marine invertebrates, there are challenges, which vary with location and even season. Salinity affects metabolism, and every marine creature must find a way of regulating salt and water balance at a cellular level. Some habitats such as the deep sea are constant, but at the mouth of an estuary, for example, salinity may fluctuate widely with each tide or with flood water. Temperature also affects metabolism. Chemical reactions typically occur faster in warmer water and very slowly in cold water. Polar species have therefore evolved special enzymes that work best at low temperatures.

Clearly not all groups or species are represented in every habitat. But there are some places that support representatives of almost all the groups and where the diversity is incredible. Coral reefs are particularly diverse because they have an

Above *Filter-feeding, barrel-like sea squirts, or sea tunicates, attached to coral. Though they are sedentary, their larvae, like those of many marine invertebrates, can swim.*

Previous page *A detail of the oral disc on top of the body of a sea anemone. The grape-like vesicles contain stinging cells.*

Opposite *A daisy brittlestar on seaweed at night. It has no head or heart but is a predator, using the many sucker-like tube feet under its arms for walking. It can shed an arm if grabbed – but can also regrow it.*

Above *A sea lemon nudibranch, or sea slug, beside purple sea urchins and spiny brittlestars. Its colour comes from, and provides camouflage on, the yellow sponges it eats, but its name derives from the defensive lemon scent it emits.*

excess of two of the three essential ingredients for growth – warmth and sunlight. The third ingredient, water-borne nutrients, is in short supply, but coral reefs have developed a community of animals and plants that make up for this by efficient recycling. The process starts with the dinoflagellates living within the corals themselves. They provide food and help to make the calcium carbonate skeleton that supports the coral polyps. In return, the coral gives the dinoflagellates both nitrogen and phosphorus as well as a safe place to live. A coral's waste products are used as nutrients by the dinoflagellates, which produce sugars through photosynthesis that are passed back to the coral. As the coral uses the sugars, it releases nutrients, which are passed back to the dinoflagellates. So the nutrients go around and around.

All other invertebrates that have a mutualistic relationship with single-celled organisms also recycle their nutrients. These include sponges, sea squirts and giant clams. Fish feeding on the reef excrete nitrogen and phosphorus as well as other nutrients, and these, too, are absorbed by plants. Nutrients are also imported into the coral ecosystem by fish that feed away from the reef but rest among the coral.

A very different assemblage of marine invertebrates is to be found offshore in the temperate region coasts where there are firm substrates for plants to cling to. Only in summer is the sun's energy strong enough for significant plant growth. But in winter, the stormy seas stir up nutrients from the seafloor. So the ecosystem is one of seasonal boom and bust. Invertebrates from such places are generally larger than their tropical counterparts. The giant Pacific octopus, for example, which lives in temperate waters, can grow to be 7m (23 feet) long.

The polar regions are even more extreme in their seasonality. For much of the year, the animals live in total darkness, and growth is on hold. When the sun reappears and the blanket of sea ice melts back, plants (phytoplankton and algae) make the most of the sunlight and released nutrients to grow and reproduce.

The plankton blooms that result are massive, and the invertebrates in the zooplankton and on the seafloor feast on the living and dying phytoplankton. The cold water slows their metabolism and therefore growth, but the animals tend to be long-lived, and many invertebrates are giants compared to their warm-water cousins.

Giants can also be found in the deep ocean, where conditions can be similar to the polar seas, with little light and extremely cold temperatures. Much of our scant knowledge of this environment is gleaned from those animals that live at depth but make forays into shallower water. One such creature is the large, aggressive Humboldt squid, which journeys to the surface at night to hunt fish.

This chapter tells the stories of the Humboldt squid and some of the other marine invertebrates that demonstrate amazing adaptations to cope with different ocean environments. They also represent the incredible success of the group: of all the known species on our planet, 97 per cent are invertebrates, and the majority either live in the sea or are descended from organisms that did.

Above *California spiny lobsters stacked in a daytime hideaway, their feeler antennae projecting out. Flanking them is purple California hydrocoral – a colonial hydroid animal with a limestone skeleton like a true coral.*

Predatory life in the fast lane

Above *Humboldt squid fighting. The loser may be eaten. These fast-moving predators have excellent sight for hunting at night and can be up to 2 metres (6.5 feet) long.*

Opposite *A squadron of night-hunting Humboldt squid, pulsing red flashes as they go. The likelihood is that they work cooperatively, communicating through complex colour signals.*

A Humboldt squid is a formidable hunter, swimming at up to 24kph (15 mph). It grabs its prey with two tentacles equipped with suckers ringed with sharp hooks and bites it repeatedly with its razor-sharp beak. Smaller fish are devoured in an instant, larger ones are quickly stripped of flesh.

Scientists believe that Humboldt squid hunt cooperatively, driving schools of sardines onto a reef or into a tight ball before moving in for the kill – behaviour filmed for *Life.* But whether the squid are acting together is hard to say. They can rapidly change colour, from deep red to white, using specialized skin cells called chromatophores. Humboldt squid have been seen flashing synchronously when fighting over prey, which indicates complex communication. And while hunting together, they flash red continuously, but no one knows whether this is just excitement or communication that helps them to herd the fish.

At night, dancing just above the surface of the Sea of Cortez, you can see hundreds of lights. They belong to boats, night-fishing for deep-sea squid. The locals call the squid *diablos rojos,* or red devils, because of their flashing red mantles and the fact that they attack the fishermen when handled on deck. They are also called jumbo flying squid or Humboldt squid, after the ocean current in which they were originally found.

Humboldt squid have a lifespan of one to four years. But in their short lives, they grow to be up to 2 metres (6.5 feet) long and about 45kg (100 pounds) in weight – an extraordinary growth rate. Tagging has shown that, by day, these squid range between 200m and 700m (665-2300 feet) deep and are very active. But how they remain active at such oxygen-depleted depths no one has yet been able to explain. At night the squid rise to the surface to feed, in schools of up to 1200. They have excellent vision for night-hunting, mostly feeding on lantern fish and sardines, but cannibalism also occurs. When a squid is hooked on a line, it's often devoured by others nearby. A quarter of squid stomachs analysed contained remains of other squid.

Accomplished predators they may be, but Humboldt squid are in turn eaten by marlin, swordfish and seals and are a staple of sperm whales. There are increasing numbers of sperm whales in the Sea of Cortez, indicating that Humboldt squid are now present in significant numbers. Some scientists believe this is because populations of tuna, swordfish, marlin, wahoo and shark have been overfished. Certainly, when longer-lived and slower-growing fish species are removed from the food web, shorter-lived, fast-growing ones such as squid move in and take over. Male Humboldt squid mature at ten months, females at a year. But females produce millions of eggs in their short lifetime, which explains why squid bounce back from overexploitation more easily than most fish do.

More recently, Humboldt squid have been found much farther north than usual, from California to British Columbia. A rise in water temperature and a decrease in fish populations are thought to be factors behind this massive range expansion, which reveals the advantage of a short, fast-growing, fast-breeding lifestyle.

A giant sacrifice

Down in the cold, dark water of the northern Pacific, under a ledge, something reddish brown can be seen moving. It's the mantle of the largest octopus species in the world, expanding and contracting as the animal breathes – sucking water in over its gills and forcing it out through its siphon. If threatened, a giant Pacific octopus can use the mechanism to jet-propel itself backwards. But for this female, starvation will be what kills her, her life sacrificed for her young.

Giant Pacific octopuses can be found down to 750m (2460 feet) around the whole of the northern Pacific rim, from California up through Alaska, the Aleutians Islands and down to Japan. Males are larger than females and can weigh up to 40kg (88 pounds), though there is one record of a 182kg (400-pound) beast. Of all the octopuses, the giant Pacific is thought to be the longest lived (though a recently discovered, deep-dwelling blue octopus may take the record). But life expectancy is still a mere three to five years, and so to breed before it dies, a giant octopus must grow fast.

When a female is sexually mature, she releases a chemical to attract males. Should two arrive together, they may fight for the chance to mate. But her chemical attractant stops them attacking and eating her – a wise precaution, as cannibalism is common among octopuses. Once she signals to a male that he is the chosen one, he uses his specially modified third right arm to insert packages of sperm (spermatophores) into her oviduct. She now has just months to live.

Her next task is to find a maternity den, preferably below 15m (50 feet) under a large rock with a small opening. She slithers into it, reaches an arm out and pulls in surrounding rocks to seal the opening. Moving to the roof of the den, she produces her eggs, one at a time, fertilizing each as it passes through her reproductive tract. Using her saliva and small suckers near her mouth, she weaves about 200 eggs into a bunch that she glues to the roof. Over about three weeks, she strings up between 20,000 and 100,000 eggs.

For the next six to seven months, the female tends her eggs. She grooms them constantly to prevent bacteria, algae and other animals such as hydroids growing on them and blows water over them to ensure a steady supply of oxygen. As the babies form, they move to the slightly wider ends of their eggs, and their large, dark eyes can be seen. The female can't leave to find food, as the eggs would quickly become prey to starfish, crabs, fish and other opportunists, and so she gradually starves. One night, the eggs begin to hatch. The female jets water over them to help the youngsters break free so they can swim to the surface under cover of darkness, when most fish are asleep. Once all have exited, the female leaves the den to die, her life sacrificed for her single brood.

The young join the plankton at the surface, feeding on larvae and other animals smaller than themselves. At just 6mm (0.2 inches), they are extremely vulnerable, and their survival rate is probably less than 1 per cent. But after 4 to 12 weeks, if they survive to reach at least 14mm (half an inch) in size, the young descend to the seabed, where they will spend the rest of their lives. Here they may fall prey to fish, seals, sea otters and even sperm whales. It will be three years before the survivors reach gianthood and are sexually mature, ready to start the whole process again.

Above *A female giant Pacific octopus guarding thousands of eggs in her den. They need constant grooming to make sure they get oxygen and that other organisms don't grow on them.*

Opposite *A giant Pacific octopus, larger than a human. When it hatched, it was the size of a rice grain, but it reached its giant size in just three years, feeding in the rich coastal waters of the North Pacific.*

The art of camouflage and seduction

Above *Male giant cuttlefish at their mating ground, jostling and assessing each other for size. A female sits in the centre.*

Opposite *A cuttlefish flashing signals. Its body colour and pattern can, in an instant, go from perfect camouflage to fast-changing colours and patterns signalling mood and intent.*

Hidden on the seabed in the southern waters of Australia lives the largest cuttlefish in the world. Like all cephalopods (cuttlefish, squid and octopuses), the Australian giant cuttlefish is short-lived, and though it grows to 1.5m (5 feet) in length and can weigh up to 13kg (29 pounds), it lives for only one or maybe two years. To get so big so fast means devoting as much energy as possible to growth. It does this by spending an astonishing 95 per cent of its life at rest. Its secret is the art of camouflage.

Avoiding detection makes the Australian giant cuttlefish ultra-efficient at finding prey and avoiding predators. It has excellent vision, by day and night, allowing it to assess the background it's against and blend in with it.

That it's well camouflaged at night suggests nocturnal predators are a big threat. Known predators include nocturnal snappers and mulloway. During the day, if dolphins swim overhead, a cuttlefish will drop to the seabed and instantly colour-camouflage itself using its chromatophores (pigment-filled elastic cells).

A giant cuttlefish has three types of colour camouflage. The 'uniform' mode, rarely used, is an even distribution of colour. The 'mottled' mode is a mix of small dark and light blotches that match the size and shape of blotches such as algal patches in the background. 'Disruptive' patterning uses varied large dark and light patches tailored to the background to disguise the animal's outline, often combined with mottling to complete the

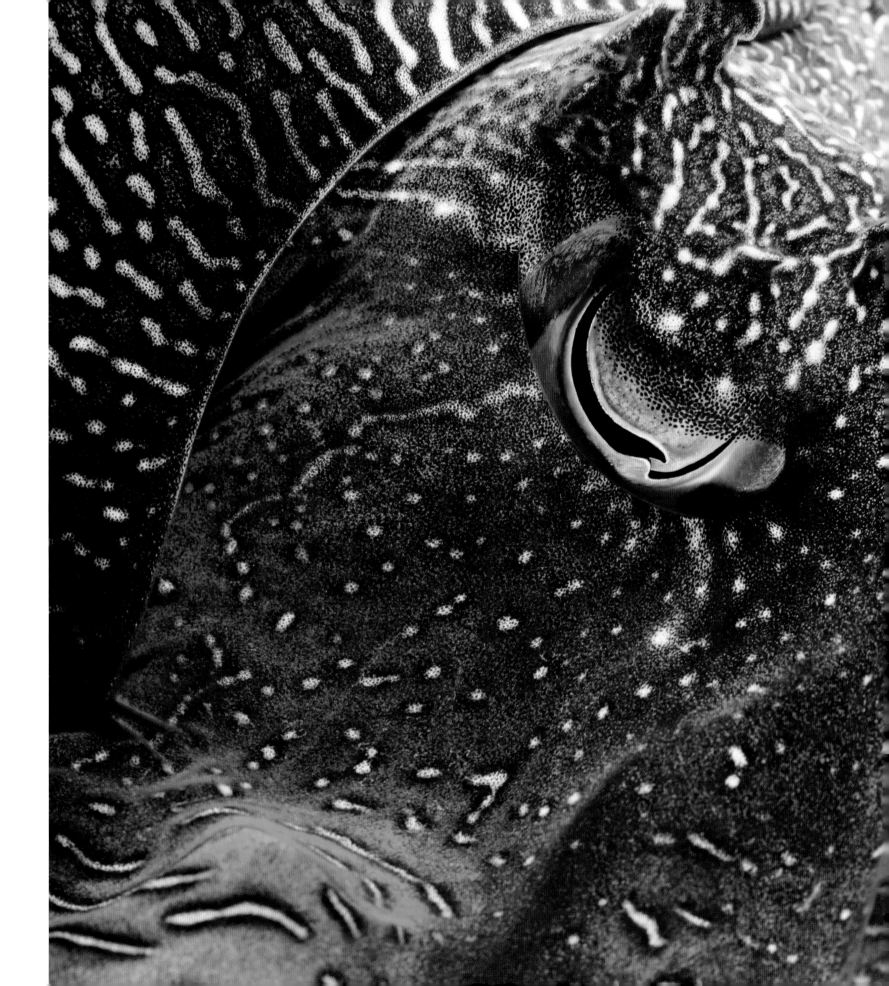

Above *A male, suffused with the colours of passion, holding a female while he inserts his fourth arm into a receptacle under her mouth and deposits sperm sacks.*

Opposite *A male cuttlefish (centre) guarding a female (bottom) against a rival. While he is engaged in a tussle, there is a risk that a small, 'sneaker' male may move in unnoticed and mate with the female.*

female simply swims off or may even bite him. If successful, a male will 'flush' the female's mouth area by squirting water from his siphon, presumably to wash away sperm from males she may have mated with. Mating itself is head to head, the male using his fourth arm to deposit sacks of sperm into a receptacle under the female's mouth. He then does his best to guard her from other males – sometimes six at once – but is seldom successful, and a female often mates with many males before laying her eggs. What is surprising is that females mate with males of any size and status, which suggests there may be no obvious benefit to being a large male or to 'flushing' the female before mating.

In fact, there is a class of small males, 'sneakers', who manage to mate with paired females despite their size. A sneaker may slowly approach a female under guard, and while the large male is fending off rivals, will quickly mate with her. Or he may hide under a rock or ledge where a female is likely to lay her eggs, and when a female explores the lair, he will mate with her in secret. The most devious strategy is to adopt the mottled coloration of a female, hide his sexual fourth arm and even mimic an egg-laying posture by holding out his arms in front, allowing him to approach a female undetected. But half the time this doesn't work because a female can still refuse to mate.

effect. Cuttlefish can also change their skin texture in an instant, raising skin papillae to produce a bumpy texture or withdraw them to produce a smooth one.

Australian giant cuttlefish are mostly solitary. But come the southern winter, thousands upon thousands migrate to the shallows of Black Point near Whyalla, putting on a mating display that even snorkellers can enjoy. They appear in May, are at their peak in early June and are gone by the end of August. The ratio of males to females is, on average, four to one, and so competition is intense. Males sort out who has the rights to a female without fighting. Those of near equal size align and stretch out their arms to assess each other. A distinctive zebra patterning moves down their bodies. If the smaller male does not move away, the larger male flares out his fourth arm as a warning. If this doesn't do the trick, he will attempt to grab the smaller male, who then usually concedes defeat.

The display to a female is short. The male uses a subtle zebra pattern on a small patch on the side of his body. Less than half of approaches are successful, and the

A female lays one large egg at a time, producing up to 40 in a day, fertilized inside her body, passed through her funnel, down her arms and then stuck under a rock or ledge or in a cave to hide them from predators such as fish. Unlike other cuttlefish, Australian giants don't then die but continue to mate and produce more eggs, which is just as well, as sea urchins seem to consume a large number of them. The eggs take 3-5 months to hatch, maturing relatively fast in the warm water. Miniature 1cm-long (0.4-inch) cuttlefish begin to hatch out in September and descend to the seafloor to hide. The adults by now have disappeared. No one knows whether they live to breed another season or whether, like other cuttlefish, they simply drift away to die.

Mass moulting and mating

From 1801 to 1803, French naturalist François Péron sailed around Australia amassing what would become the most significant Australian natural history collection ever made – some 100,000 specimens. Off Tasmania, he noted 'spider crabs, which delight in filth and mud, abounded to excess on every point in the channel.' Even today, in late autumn or winter, the channel is one of the best places to find massive aggregations of the Australian majid spider crab.

Surprisingly little is known about the species. It's found down to 800m (2625 feet), mainly on sand or silt, from which it plucks small marine creatures and algae with its pincers. Aggregations occur when the crabs migrate to shallower water to mate, creating huge piles as they climb over each other. Like all crabs and other crustaceans, the spider crab's growth is constrained by its exoskeleton (shell), and the only way to expand is to grow a larger, soft shell under the existing one and then split open and crawl out of the hard one. Calcium is absorbed into its blood from the old shell while the

soft one forms underneath, and a crab may even grow back a lost claw. Finally the crab sits tight while its new exoskeleton hardens. This is when it's most vulnerable, but it is also when the females are ready to mate.

Receptive females attract males in vast numbers, and mounds build up, some ten-deep. These may include crabs about to moult, newly moulted as well as mating crabs and shells. Mating takes place face to face and belly to belly, the male fertilizing the female internally. She will lay thousands of eggs within days, brooding them under her abdomen, tucked under her shell.

Obviously, getting together in such large numbers is one sure way to find a mate, but aggregations of hard-shelled spider crabs may also occur when the females are not receptive, and when males are soft-shelled and can't inseminate females. So it's likely that aggregations provide safety in numbers, especially when crabs are soft-shelled and vulnerable to predators such as rays.

Far left *Getting ready to moult. The only way a spider crab can grow is to split out of its old exoskeleton. During the moult, a crab may even grow back a lost claw, though this is never as large as the original.*

Left *Emerging. As the old exoskeleton splits, the crab pulls itself out, quickly absorbing water. But the new shell is soft, and its muscles are too bendy for it to walk properly, making it vulnerable to predators.*

Below *A huge manta ray sweeping over the mass of crabs, preying on those newly emerged and still soft-bodied.*

Strategies for ice life

Under the ice in McMurdo Sound, in Antarctica's Ross Sea, exists an ancient, isolated and utterly unique community. As in other parts of Antarctica, the sea is frozen over for most of the year. Only towards the end of the short summer does the icy blanket covering the Ross Sea break up, but then just for a few weeks.

Antarctic invertebrates – from sponges and starfish to corals and crabs – have had a long time to evolve here. When the Antarctic continent drifted south and finally separated from South America some 25 million years ago, the Southern Ocean was free to circulate around the continent without interruption. Gradually the Antarctic Circumpolar Current grew in strength and created a barrier between the relatively warm, northern waters and the cold southern waters. The biological isolation of Antarctica had begun.

In McMurdo Sound, little diatoms, flagellates, copepods and amphipods abound. For most of the year, they feed on bacteria in the pockets and channels that riddle the under-surface of the ice and on clumps of platelet ice

that grow down from it. Platelet ice forms in spring when supercooled water from under the Ross Ice Shelf flows into McMurdo Sound. Though lifeless, this water has a profound influence. Ice crystals form in it and gather under the sea ice in a confused layer of thin ice shards, adding an extraordinary amount of surface area on which the sea-ice community can grow. If the platelet layer is too thin one year, then the living space is restricted; if it is too thick, then there is too little water movement to replenish essential nutrients. The ideal thickness seems to be half a metre (1.6 feet). Ice crystals in the water also form 'anchor ice' that carpets the seabed up to 30m (98 feet); below that, the pressure is too great for ice crystals to form. It scours the seabed, and chunks float up to the sea ice and merge with the platelet ice, taking with it any animals unable to escape its grip. Thus, by and large the only animals found on the seafloor in these areas are those that can move, such as urchins, starfish, worms, isopods and fish.

At around 15-30m (49-98 feet), anemones, soft corals and even some sponges form the community, but below

Below A view over the sea ice of McMurdo Sound, with the volcanic Mount Erebus smoking in the background. Where the sea ice meets the shore ice, a pressure ridge has formed. Under that ice is an equally spectacular vista – of a rich and seldom-seen marine environment.

Above *Huge filter-feeding volcano sponges, surrounded by animals including anemones, sea urchins and soft corals. Other animals are also hiding inside them. In summer, warm water sweeping under the ice reaches McMurdo Sound, producing a plankton bloom that feeds the community. But growth is still very slow, and these sponges may be hundreds of years old.*

Right *Antarctic sea anemones illuminated by light from a crack in the sea ice. They are voracious predators capable of eating large animals, including jellyfish. They can move and so avoid anchor ice (here seen lying on the seafloor behind them).*

30m, the diversity is astounding. Sponges, which are slow-growing and so could not survive anchor-ice formation, dominate, creating a three-dimensional habitat for other animals. Hydroids grow from their tops and sides; featherduster worms use them as feeding platforms; fish hide in them and may lay their eggs in them; starfish and nudibranchs (sea slugs) eat them.

As spring turns into summer and the blanket of sea ice begins to melt, photosynthesis accelerates and algae multiply, covering the surface of the platelet ice. Though the brown slime cuts down the light reaching the phytoplankton below, this community of tiny light-dependent organisms now flourishes. A polynya – an area of permanently open water surrounded by ice – sits on the northern edge of the Ross Ice Shelf. Each year it gradually opens up, and sun-warmed water sweeps around Ross Island, reaching the eastern side of McMurdo Sound in mid-November. This replaces the supercooled water, resulting in a plankton bloom under the now-thinning ice. Platelet ice melting in the warmer water exposes its inhabitants, freeing yet more food.

Right *Red starfish, sea urchins and a brown mat of diatoms carpeting the seafloor under sea ice in McMurdo Sound. Starfish, along with sea urchins, dominate this area, eating almost anything, including other starfish.*

Much of this biological matter sinks, along with the plankton, and the seafloor becomes alive with activity, though in water that's near freezing, activity is a relative term – animals move slowly, almost imperceptibly.

Very different habitats exist within small distances in McMurdo Sound. In the east, southerly winter winds carrying snow dump their load on the windward side of the peninsulas of Ross Island, leaving the leeward side snow-free. This allows the spring sunlight to penetrate into the water earlier than in snow-covered ice areas.

Where this corresponds with platelet-free sea ice – places where supercooled water from under the ice shelf hasn't reached – a lot of light penetrates the water, and bottom-dwelling diatoms multiply, carpeting the seabed with brown mats. This provides food for the Antarctic sea urchin – common everywhere – which mainly eats algae but also consumes animals, from diatoms and amphipods to sponges and worms.

Seaweed also grows here – one species in the shallows and another between 10m and 15m (33-98 feet). They

Right *A mass of starfish and proboscis worms (nemerteans) homing in on and consuming a dead seal. The giant worms grow up to 2 metres long and can smell food over long distances.*

contain toxins that sea urchins can't tolerate, but the sea urchins make use of them, biting off pieces and sticking them to their spines. This creates a coat of seaweed against predatory anemones, which withdraw their tentacles if they touch the algae. Borrowing someone else's toxins is a cheap defence option. Nudibranchs make use of stinging cells from the soft corals. And one small, free-swimming amphipod places a small floating sea slug with flapping 'wings' on its back as a defence against fish, which find the 'sea angel' distasteful.

A notable feature of McMurdo Sound's community is the lack of juvenile invertebrates. This is probably due to the omnipresent *Odontaster* starfish, which eats everything, including seal faeces, dead seals, sponges and other starfish. Its dominance, along with that of the Antarctic sea urchin, is in part a result of its reproduction strategy. Rather than hiding a few yolk-rich eggs on the seabed or brooding its larvae, it releases numerous eggs and sperm into the water in late winter. This means the young are not immediately consumed by active starfish or sea urchins, and have a head-start, eating bacteria and then moving on to algae in summer.

On the far western side of the Sound, a completely different community of animals exists in Explorers' Cove, despite hostile conditions. Here the supercooled water from the Ross Ice Shelf exerts its influence through the year. The ice doesn't melt, except around

the shore, where there is freshwater runoff, and builds up to be more than 5m (16 feet) thick. The platelet layer hanging beneath it can be 3m (10 feet) thick in places, reducing any nutrient-mixing that might occur. No light can get through, and productivity is minimal. Even where the summer water from the Ross Sea penetrates, it carries no food: it has all been consumed already.

Here the seafloor consists of extremely fine sediment, accumulated over millennia of glacial run-off. This is the perfect habitat for scallops, which live here in their

Above *Proboscis worms and starfish in a slow-motion feeding frenzy on seal faeces. Nothing goes to waste under the ice.*

thousands. In the shallows, where melting occurs each summer, as many as 85 scallops per square metre can be found. In deeper water, the density drops off, but 20 scallops per square metre can be found as deep as 30 metres. There seems to be little keeping the population in check, as even starfish are rarely seen here. But as in the deep ocean, growth is slow, and the scallops remain comparatively small, because the extreme cold makes it difficult for them to precipitate calcium carbonate from the water and their shells are thin and fragile.

Animals typical of the depths occur, including pencil urchins, brittlestars and foraminiferans. The latter are unicellular animals common in shallow, warm waters worldwide and usually microscopic, but here they can reach a centimetre across. One species is predatory and appears to consume most of the settling larvae of other invertebrates, including those of scallops.

Another unique invertebrate living in this extraordinary world of Explorers' Cove is a giant soft coral, *Gersemia,*

standing some 1.5 metres (nearly 5 feet) tall. Its polyps feed from the water, collecting passing organic matter and tiny invertebrates. But in this place of negligible water movement, there is insufficient passing food to sustain the colony. So it sweeps the surrounding seabed clean of prey. To do this, it changes the hydrostatic pressure on one side of its central stalk, causing it to bend until its polyps are in contact with the sediment. After the polyps have trapped all they can, the coral straightens up and bends down to cover a different patch, and so on until it has swept the entire area in a circle around itself. More remarkable still is what follows. The soft coral, far from being attached for life to a chosen substrate, detaches itself from the rock or scallop it was using as a base and crawls, worm-like, along the seabed to another likely feeding patch.

Extreme conditions call for extreme measures, and in the cold depths are more examples of extraordinary strategies than in any other habitat, most of which we have yet to learn about.

Coral: life at its busiest

Above *A tall, tree-like soft coral feeding at night, its hundreds of tiny polyps open, with tentacles extended, filtering plankton from the water.*

Right *A reef composed of hard corals – the reef builders – with skeletons of rock-hard calcium carbonate. Like plants, they compete for space and light, taking many shapes, represented here by tabletop, staghorn and cauliflower coral. Light enables the tiny organisms (dinoflagellates) inside the coral polyps to photosynthesize and produce sugars and oxygen, which the coral then uses.*

There is one family of marine invertebrates that has created structures so immense they are geological features in their own right and provide a three-dimensional framework for communities that are as rich in species and complexity as those of tropical forests. These are the corals, found where land and ocean meet in the warm seas that stretch around the planet in a wide belt on either side of the equator.

Coral animals require specific conditions to thrive. They need warm water, ideally between 18°C (64°F) and 30°C (86°F). So if a cold-water current runs into a tropical area – such as the Humboldt Current sweeping through the Galapagos Islands – coral growth is poor. Equally, if the water becomes too warm, the corals die. But where warm currents penetrate colder water – as in Bermuda, where the Gulf Stream runs around the island – reefs thrive. Corals also need a hard substrate to grow on and access to light. So the seafloor must be shallow and the water must not carry much sediment or be diluted by fresh water running off the land.

Corals reefs are celebrated as one of the most beautiful, rich and complex communities of living organisms on the planet. Many species living together have given rise to countless interactions, both beneficial and harmful. Corals need sunlight to grow and therefore, just like plants, they compete with one another for space in the light. Some grow fast, stretching up and then spreading out, shading the slower-growing species. Others try to dominate by sending out especially long sweeper tentacles covered in stinging cells – nematocysts – to attack neighbouring colonies. Yet others attack by extruding gut filaments (mesenterial filaments) to digest, quite literally, their closest neighbours. Typically, the slower-growing ones tend to rely on more aggressive tactics to create space for themselves.

Many of the vast number of species living on a reef have evolved symbiotic relationships where one or other of the partners benefit, sometimes both. Corals themselves exist through mutually beneficial partnerships. Each

coral polyp nurtures dinoflagellates – single-celled organisms, which use sunlight to photosynthesize and produce sugars and oxygen that the coral uses. In turn, the coral produces carbon dioxide, nutrients and a safe home. Without dinoflagellates, corals could not deposit calcium carbonate at the rate they do and form the skeleton of the reef. Other invertebrates, including sea anemones, snails and giant clams, also give dinoflagellates a home. Indeed, without these plants, giant clams wouldn't be able to grow such huge shells.

Reef crustaceans (crabs, shrimps and the like) also go in for relationships with a wide variety of animals, from corals, anemones and sponges to molluscs and echinoderms. Such relationships enable many species to coexist in a small area, but not all are beneficial to both parties. The smaller of the two could be a predator, parasite or a scavenger, feeding on the dead tissue of the host, or it could merely be seeking protection.

The fire urchin, found in the Indo-Pacific, has many small, short spines that inflict a great deal of pain on divers unfortunate enough to brush against them. But its stinging spines provide protection for a number of creatures, including a shrimp, an urchin crab and a cardinal fish. Carrier crabs will even carry a fire urchin or a jellyfish on their backs as security against predators.

Small, almost translucent and nocturnal, and therefore rarely seen, the fairy crab grows tiny hydroids – tree-like colonial animals with stinging nematocysts – on its back and legs. These provide a service more important than protection: they collect plankton animals from the water, the bigger of which the crab appropriates.

One well-known relationship is that between cleaner shrimps and their clients, typically fish. Many fish are attracted to the dancing and swaying antics of cleaner shrimps, which position themselves in conspicuous places on the reef. Usually found in pairs, they are visited by a succession of fish. A shrimp feeds both on its host's mucus and on any small parasites, working its way over the body, nibbling here and there as it goes.

These are just a few of the multitudinous relationships in the packed reef communities. The interdependence of marine creatures in this habitat is second to none, and we have only just started cataloguing them, much less understanding them and the extraordinary strategies that have evolved in the cities of coral.

Clockwise from opposite
*Close-up details of coral on the
Great Barrier Reef – whip coral
with its polyps extended,
lace coral and green scroll coral.*

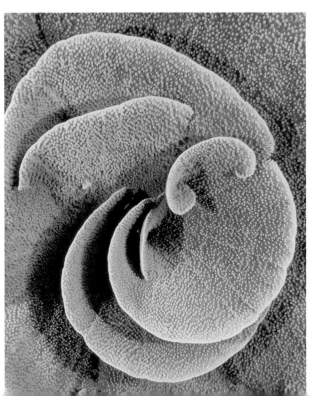

chapter 2

Fabulous fish

THE PLANET'S MOST DIVERSE and widely distributed class of vertebrates are the fishes. From mountain streams to the abyssal depths, they occur around the globe in virtually every watery habitat. To date, there are 28,000 or so known species (compared to just 5400 of mammals), with doubtless many more waiting to be found.

The secret of their success is their basic layout – a backbone, gills, a hinged jaw and fins – which evolved more than 400 million years ago. The huge advantage of developing hinged jaws and paired fins all that time ago was twofold: hinged jaws allowed a wider range of food to be eaten – allowed a fish to be a predator rather than a bottom-detritus feeder – and probably made respiration more efficient, since being able to gulp oxygen-rich water meant fish were free from a lifetime of forever having to swim forward to breathe. Fins combined with a flexible backbone meant that propulsion was not only more efficient but also that power was more controllable.

The dynamics of how a fish swims is extraordinary. Water is denser than air, and drag is a major obstacle to efficient movement. The best body shape for moving fast through open water is the streamlined shape that evolved in fish such as sharks and, later, the fastest swimming fish of all, such as tuna and billfish, combined with a propulsion system of solid muscle.

But speed is not everything. Depending on where a fish lives and what it eats, precise movement is often more important, enabled through fins used in different combinations and different ways. A flying fish has elongated pelvic fins that can be transformed into wings if it needs to make a quick getaway, while the Australian weedy seadragon has tiny fins on its neck and back that it vibrates rapidly, like a miniaturized helicopter, to manoeuvre itself precisely among the seagrass and kelp where it lives, with no apparent movement of its body. The development of a swim bladder in modern fish gave them a gas-filled chamber that allows them to adjust their body density to match that of the surrounding water so, when they are not swimming, they neither sink nor pop up to the surface, unless they want to.

No aquatic region is off-limits, whether salt water or fresh water – even the tops of massive waterfalls have been colonized by gobies using specially adapted fins as climbing aids. Other fish such as the mudskippers have colonized tidal mudflats by transforming fins into walking aids and developing ways to get oxygen while out of water.

The stories in this chapter touch on just some of the wondrous behaviours that fish have devised in their exploitation of most of the water-covered surface of the planet. But the key to our fascination with the fish is not just that they are so varied in looks and behaviour but that they also inhabit a realm that we can enter only briefly and still know so little about.

Previous page *A streamlined, muscular great white shark displaying a classic and ancient fish shape. It swims bending its body from side to side in a shallow curve – a style that can achieve great speed in a very energy-efficient way for its body size and shape. Special skin reduces the drag as it hurtles through the water.*

Above *A bignose unicornfish with a small mouth for picking off algae and small animals. Colour is used for display and camouflage (it can change to a muddy brown when sleeping).*

Opposite *A striped marlin feeding on Pacific sardines. It is one of the fastest fish in the sea, with a rapier shape powered by special oxygen-efficient muscles.*

Big-mouth and the snappers

Right *A racing pack of dog snappers on the Belize Barrier Reef, spiralling upwards. The pack is composed of males pursuing females, who are swimming upwards to release their eggs near the surface. They do this in the evening after the full moon.*

Opposite *Cubera snappers releasing a milky cloud of eggs and sperm at the peak of the spawning rush off Gladden Spit.*

Gladden Spit lies between the shallow waters of coastal Belize and the deep waters of the Caribbean. Inside the reef, flights of stingray glide past, and shoals of reef squid hang in the water like platoons of spaceships on standby. In the seagrass beds live a great variety of marine creatures, together with many young fish that choose to use the seagrass as a nursery until they are big enough to venture out into the deep water.

Heading out to sea from the beautiful emerald waters of this nursery zone, the waves suddenly become bigger and the swell more powerful, and things feel a lot more serious. When the sunlight shafts down and vanishes into the dark, you feel that the deep blue sea below the boat is bottomless. But the water is far from lifeless. Gathering 60m (200 feet) or so below are the participants

in one of the most impressive of fish events in the Caribbean. From March to June, in the days leading up to the full moon, huge shoals of snappers congregate here to spawn. The shoals, each made up of thousands of individuals, are segregated by species – dog, mutton and cubera snapper come together with their own kind. At dusk, around the full moon, the fish can be found in just 30m (98 feet) of water. On some hidden signal, members of the shoals suddenly become animated, and the action begins. Groups of hundreds of fish burst from a main shoal and spiral upwards in racing packs.

The females are making a burst to the surface, each closely followed by males hoping to add their sperm to the mix when she releases her eggs. As a female races upwards, the expansion of her swim bladder

may help force the eggs from her body. In any event, as she gets to within 15m (50 feet) of the surface, she releases hundreds and thousands of eggs, and the males respond with their sperm, rapidly turning the water milky white. Oils released with the sperm cause the water surface over an area the size of a tennis court to become smooth and flat – a counterpoint to the action below. Suddenly, an enormous shape can be seen moving up through the cloudy water. It's the largest fish in the world, but it's here for the eggs, not the fish.

Whale sharks are filter-feeders and normally feed on phytoplankton and krill, but every year, large numbers of these leviathans gather at Gladden Spit for the snapper spawning. They don't need to be swimming forward to feed, and when consuming the spawn, they often hang vertically, taking in great gulps of water, each mouthful marking the end of millions of eggs.

Every evening for some ten days after the full moon, this spectacle will be repeated. At the start, the females are fat with eggs, but as the days pass, they get thinner and the action gets less frantic until it finally peters out. The snappers probably wait for the full moon because the bigger tides mean the eggs will be dispersed farther. In ten days, millions if not billions of eggs are released by these fish – far too many for the sharks and other predators to consume, making sure that millions survive. But they face countless other challenges as they drift the oceans, and only relatively few will grow to adulthood to take their place in the incredible spawning and feeding spectacle of Gladden Spit.

The long climb to fish nirvana

The islands of the Hawaiian archipelago are the remotest in the world. Lying some 3862km (2400 miles) from the nearest continental shore, they are still driven upwards by the volcanic forces that originally thrust them high above the waters of the Pacific. The river systems on these islands are both short and steep, and waterfalls plunge from the cliffs into the sea – a curtain of water linking the fresh water to the ocean.

The archipelago is so remote and has, in geological terms, been around for such a short time that there has been little chance for life to colonize its freshwater habitats. But what the streams lack in numbers of species is made up for by the dramatic lifestyles of its inhabitants. Four of only five species of native fish are gobies. Their pelvic fins are fused into a disc with which the little fish can literally stick themselves onto rocks, and these discs play a critical role in their early lives.

After hatching, the larval gobies travel downstream and out to sea to form part of the plankton soup, feeding and growing over months until they are about 10-25mm (0.4-1 inch) long. Then they swim back to the islands, where they will spend their adult lives. Initially it's not difficult for the small fish to work their way up the lower reaches of the streams, but inland they are abruptly faced with a Herculean task – the monumental Hawaiian waterfalls, some of which are more than 122m (400 feet) high. This is when the discs come into play.

The gobies gather at the edges of the waterfalls, where the water forms a consistent film or rivulet over the rocks. Suddenly one fish will start to climb, and this is the cue for all of them to head upwards. There are different strategies. Some gobies edge up gradually. Others flick themselves out of the water and land on the rock face – sticking on with their pelvic discs. For all of them, it's the start of an incredible journey.

They must now employ climbing techniques. Some use a combination of mouth and disc to inch their way up the rock face in a caterpillar-like motion. This slow, steady method allows a fish to keep climbing for substantial distances between rest stops. When it does need a break, it finds a pocket of calmer water or an indentation in the rock. Other gobies have a far more flamboyant way of climbing. They have enlarged pectoral fins, which they use like a swimmer doing the butterfly stroke to power up through the water, aided by flicks of their tails and indeed their whole bodies. These gobies tend to climb in shorter, faster bursts than the caterpillar climbers.

If you stand at the top of one of these waterfalls, the drop below seems to go down for ever, and the force of the cascade of water bouncing from the rocks generates a thundering roar. Who knows how many of the thousands of gobies that start the climb make it to the top? But enough must for the species to survive. So why then go to all this effort? Above the waterfalls is something approaching a fish nirvana: a breeding place with few predators and little competition. It's a classic example of how fish will find and exploit a niche, no matter what they need to go through in order to reach it.

Below An o'opu rock-climbing goby on its marathon ascent up the cliff from the sea. On its underside, fused pelvic fins form a sucker-like disc, which it uses to cling to the rock as it makes its way up the waterfall.

Opposite The goby taking a rest during the slow, steady climb up the rockface, using a combination of mouth and disc. Its reward will be a breeding place at the top, comparatively free of predators.

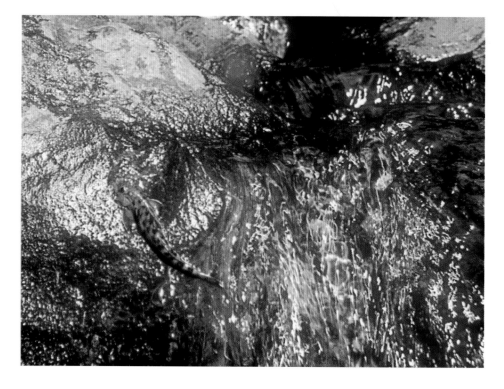

Precious eggs and caring fathers

The ability to exploit every sort of marine habitat is one of the great strengths of fish as a group. This includes making use of man-made structures. Take for instance the warm shallow waters around a pier off the south coast of Australia. Here live seahorses, squid, puffer fish, rays and stargazers. Shoals of tiny fish flit in and out of the protective cover of the piles. Even great white sharks have been seen – possibly cruising here when the local seal colony is proving less productive.

Around the pier are seagrass beds, which provide cover for many forms of life, including a fairy-tale creation of a fish, the weedy seadragon. It has a very unfishlike appearance, and its fins flutter at such a rate that it moves without obvious means of propulsion, though slowly, which helps to camouflage it among the seaweed and seagrass fronds that it so closely resembles. It feeds mainly on tiny crustaceans such as mysid shrimps. A weedy will hover close to a swarm and hoover up shrimps one by one. Several weedies may be attracted to a swarm, but gatherings of weedies usually occur only in October and November, when the mating pairs form.

The prelude to mating is a dance, which starts as the light fades at the end of the day, each fish mirroring the moves of the other. As night falls, the pair drift off into the darkness. No one has witnessed exactly what happens, but instead of releasing hundreds if not thousands of eggs into the sea, weedy seadragons choose to guard their eggs 24 hours a day, not in a pouch, as seahorses do, but in spongy tissue on the male's tail. This restricts the number of eggs that can be cared for to the number of 'egg cups', about 120, that appear on the brood-patch tissue which develops prior to mating. By morning, he will have row after row of purple eggs attached to his tail. For the next few hours, he appears to try to straighten out his tail to get the eggs all lined up, swimming slightly over to one side, as if off-balanced by his precious load.

The eggs mature on the brood patch for about a month, sometimes gaining camouflage from filaments of weed that grow on them. When they are ready to hatch, the male shakes his tail to help release the miniature weedy seadragons so they can swim off to take their own chances in the coastal waters of southern Australia.

Left *A male weedy seadragon, carrying eggs, which have been deposited in egg cups on the spongy brood tissue on his tail. The eggs are in the process of hatching, and he is shaking his tail to release them.*

Opposite *Newly emerged baby seadragons. They are on their own from the moment they hatch but have emergency supplies of food in the form of yolk sacs.*

Next page *A weedy seadragon male with a fresh brood of eggs, receiving oxygen from blood vessels supplying his patch. Carrying the eggs with him is effectively a 24-hour guard, and he relies on his weedy camouflage to avoid being spotted by predators.*

Air-gulping, mud-sucking champions

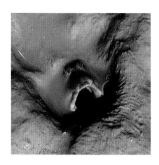

Above *A male Japanese mudskipper in his burrow, gulping air to take back down to the nursery chamber.*

Below *A male, fins erect, performing a waggling walk on his pectoral 'walking aids' to attract a mate to his mud burrow. He provides total parental care for the resulting eggs, from making a special incubation room for them to oxygenating them and making sure that young fish hatch and leave home on the right night.*

Opposite *The much larger blue-spotted mudskipper with his more flamboyant fins, jumping both to proclaim ownership of his patch of mudflat, covered in nutritious diatoms, and to attract a female to his burrow. If challenged by another male, he will fight for his prime breeding territory.*

There can be few tougher places for a fish to live than on a tidal mudflat. Extreme changes in salinity and mud movement are major problems, but the biggest challenge is lack of water for substantial periods. Mudskippers have been able to exploit this rich niche by adopting an amphibious lifestyle, but it has required major changes. Their skin helps them to breath air, but they have also adapted their gills to hold water and lock shut when on the mud. In addition, they have turned their pectoral fins into walking aids, like crutches. Their reward is food, from small invertebrates, such as crabs and flies, to algae and microscopic life in the mud, such as diatoms, and there is very little competition for it.

Much is still to be learnt about how these extraordinary fish survive such harsh conditions, but recent studies of the Japanese mudskipper *Periophthalmus modestus* have given us some fascinating insights. Like most mudskippers, it uses its mouth to excavate a burrow, where it sits in the heat of the day at low tide, avoiding predators and the worst of the sun. Most important, the burrow is a safe place for eggs to be incubated, avoiding releasing them into a dangerous sea. A male generally digs a J-shaped tunnel, with the tip of the J some 20cm (8 inches) below the surface. But the water here is low in oxygen, and eggs critically need a good oxygen supply to develop and hatch. To get round this problem, the male does something ingenious. Once he has lured a female to his burrow, she will lay her eggs on the walls of the upturned end, where he fertilizes them. He then becomes their carer. Key to his success is the pocket of air at the end of the J in which the eggs sit. But this air pocket won't last long. So the male collects air, gulping it down at the mouth of the burrow at low tide and then swimming the length of the tunnel to release it in the chamber.

After six or seven days, the eggs are ready to hatch. But they need to do so under cover of darkness, when fewer predators are around. To induce hatching at the right time, the male waits for a high tide at night and then gulps the air out of the chamber so that it floods. Once they are bathed in seawater, the eggs hatch and the next generation of mudskippers goes on its way.

Floating and flying away

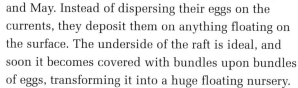

Above and below Mirror-wing flying fish powering out of the water and, with fins expanded, gliding to safety. Pursuing the shoal are predatory dorado – fast swimmers but no match for the flying mirror-wings.

Opposite A mass of flying fish spawning under a raft of palm fronds, the males releasing their milky sperm to fertilize the eggs that the females are attaching to the fronds above. The raft will become a floating nursery.

A raft of matted palm fronds and debris floats on the surface 48km (30 miles) off the shore of Tobago in the Caribbean. Having been adrift in the sea for months, it has already become home to a range of creatures – a place of refuge in a huge ocean. But now new animals have arrived to make use of this raft, as a nursery for their eggs.

Thousands upon thousands of flying fish have gathered to spawn. Activity is frenetic, and soon there is a huge mass of fish under this one small raft. In the Caribbean, flying fish gather to spawn between January and May. Instead of dispersing their eggs on the currents, they deposit them on anything floating on the surface. The underside of the raft is ideal, and soon it becomes covered with bundles upon bundles of eggs, transforming it into a huge floating nursery.

But all this action doesn't go unnoticed. Hovering on the periphery are dorado, or dolphinfish. They give an ominous impression of muscle and power, with a dorsal fin that runs from just behind the head down the whole length of the back and a sickle tail signifying speed. And speed is what they need if they are to stand any chance of catching the spawning flying fish.

Racing in, a dorado tries to single out a likely victim. But in an instant, the flying fish fires forward, racing just under the water's surface with powerful flicks of its tail – and vanishes. Above the surface, it's now visible – and flying.

With its elongated pectoral fins spread like wings, it leaves the dorados far behind under the water. As its momentum slows, it glides towards the water's surface and then propels itself upwards with a few more powerful strokes of its tail – expanded, flat pelvic fins functioning as stabilizers. In this way, the flying fish is able to glide for up to 50m (164 feet), taking it well beyond the reach of its predator – an original escape mechanism for a water-bound fish.

The strange life of the convict family

Above A parent convict fish beginning the day's never-ending chore of clearing sand out of the burrow with her mouth. The infants, meanwhile, are swimming off to feed.

Below Young convict fish exiting. They won't return until dusk, using disguise as protection.

Opposite A shoal of young convict fish feeding on microscopic marine life. Their parents never leave the family home. So what do they eat? Could it be their young?

In the world of fish, there can be few stranger examples of family life than that of the convict fish of the southwestern Pacific. Adult convict fish are roughly 50cm (20 inches) long and eel-like in shape and spend virtually all their lives in the burrows they dig around the coral reefs. The extraordinary thing is that a pair shares its home with thousands of their young, which lead a completely different lifestyle, leaving the home daily to feed and moving as if one giant organism.

The burrow may have up to four entrances, each marked by a fan of sandy material. A few minutes observation will show that the fan is created by a parent emerging from the hole and spitting out sand and bits of coral – the residue of diligent enlarging and cleaning. During daylight, housekeeping is almost non-stop, as sand is washed into the hole by tides and currents. In a single day as much as 3 kilos (6.6 pounds) of sand might be collected and spat out of the hole by the parents.

At dawn, the first of the youngsters will appear at a burrow entrance. Instead of the adult blotchy patterns,

it has two black stripes running head to tail. It's soon joined by another and then another youngster, turning into a stream of tens and hundreds and even thousands. The snake-like shoal that heads off into the open water can be up to a metre (more than 3 feet) wide.

The young fish feed on plankton on the open reef, moving as one – sometimes forming a rolling ball, sometimes a bulky shape. This is an anti-predator defence, but there's also an insurance policy. They resemble another fish species that lives in the same area and also travels in shoals – the striped catfish – which is poisonous and so tends to be left alone.

Having spent the day feeding on the reef, the shoal returns to the burrow at dusk, racing back into the hole, like water down a plughole. At night, the young fish rest suspended from the ceilings of the tunnels, hanging by their heads from threads of mucus. Indeed, the walls of the tunnels are lined with the remains of old mucus.

Clearly the young fish benefit from their parents' labour by the security of a well-maintained burrow to return to. But it is not just they who benefit. One of the great mysteries about convict fish is what the adults feed on, since they never leave their burrows. They may get food from all the sand and coral they shift during the course of the day – small invertebrates and microfauna, for example – but examination of their stomach contents have failed to show any evidence of this. The adults might be feeding on the faeces of the young fish or even on the mucus they produce, or the young may be regurgitating food for the adults. Could the adults even be eating some of their young?

There is still much to be discovered about the lives of these and so many other types of fish. Because we can spend only limited periods of time under water, our opportunities to study what really goes on are cut short. But this is what maintains our fascination with life under the waves – there is still so much waiting to be discovered.

chapter **3**

Irrepressible plants

PLANTS ARE CONSTANTLY BATTLING – with each other for resources and mates, and against predators – just like animals. Though they may collaborate with each other, more often they deceive and parasitize, and in some cases they even hunt. Why we fail to notice such dramatic behaviour is because, first, we tend to consider plants as rooted to the ground and therefore literally inanimate, and second, plant behaviour happens so slowly that we seldom notice it. In fact, it's that slow pace of life that is the key both to their success and to understanding them. And if you shift perspective to a plant's point of view, a world of complex and wonderfully sophisticated behaviour comes into focus.

Like all living things, plants must compete for water and nutrients, but their greatest struggle is for light. Without light, there is no photosynthesis and therefore no growth, and so plants will do anything to get as much sunshine as possible. A young sunflower seedling, for example, turns to and fro scanning the horizon for the position of the rising sun and fixes its growth to face east to gather as much light as possible. It's in this struggle for light that plants exhibit their most hostile characteristics and vigorous adaptations. Climbers will exploit other plants, clambering up their stems or trunks, using lassos, suckers and even strangling coils in their efforts to reach the sun.

Plants use light to measure the passage of time. They can see the coming of spring, of winter or of the dry season and so can decide when to flower, when to set seed or when to shut down. They are masters of timing. This, combined with their sense of touch – the tips of some tendrils are far more sensitive than even a human hand – can turn them into active aggressors against animals. The Venus flytrap can even count, using highly sensitive hairs to calculate when to close its trap.

Plants are also clever manipulators. The relationship between them and their animal pollinators may appear balanced, but look closer and you'll see that the plant usually has the upper hand. For a start, it controls the amount of nectar it produces: too much, and the pollinator may become satiated and won't bother to visit other plants; too little, and it may give up visiting. If the plant gets it right, it has the pollinator rushing from flower to flower (of course, carrying pollen as it goes) receiving just enough nectar to fuel its visits and just enough on top to allow it to survive.

Plants have colonized every land habitat, including some where no animals survive. They have been residents on land far longer than animals, with a family tree stretching back nearly half a billion years. Today they include the biggest, tallest, oldest and, if you exclude bacteria, the most numerous living things on the planet. Every land animal depends on them for survival, whether directly or indirectly, and in the end, all flesh is indeed grass. So rather than helpless victims simply providing fodder for animals, plants are often the masters of animals, ruthlessly manipulating them to their will.

Above *An insect's view from inside a rafflesia flower dome in Sabah, Borneo. The giant flower – the world's largest – is a trap, attracting carrion-loving beetles and flies with the smell of rotting meat, holding them prisoner and only releasing them when they have completed their job of pollination.*

Right *Sunflowers in southern France, all facing the sun to gather as much of its energy as possible, for both photosynthesis and the ripening of their seeds.*

Previous page *An ancient oak grove sculpted by wind and rain, photographed on 1 January, on Dartmoor, England. Ferns, mosses, liverworts and lichens cover every tree-trunk, branch and rock.*

Longevity as a solution to adversity

Right *A bristlecone pine in the White Mountains of California – several thousand years old and one of the longest-living things on the planet. The climate here is so harsh that few other plants can survive, and so there is little competition. But extreme cold, wind and drought mean that the bristlecones grow very, very slowly.*

At sunrise on midwinter's day in the White Mountains of the western USA, a lonely tree casts the longest shadow of the year across the snow. This bristlecone pine has seen many a midwinter's day, 4740 of them in fact, being the oldest living tree on the planet. It was a seedling before the Egyptians laid the first stone of the pyramids and was reaching maturity before the birth of Christ. With its ancient fellows, it lives above 3048 metres (10,000 feet) in the Eastern Sierra of California. It's a harsh place: very cold, very dry and with thin, alkaline soil. It's so harsh that the bristlecone has the place almost entirely to itself, a fact that appears to be critical to its survival.

Time for a bristlecone pine takes on a different dimension. To put on 2cm of girth can take a hundred years. The biggest and tallest can reach more than 18 metres (60 feet) in height. These trees tend to live in the most favourable locations but often die young, at the tender age of a millennium and a half. The *really* old trees tend to live in the harsher sites, with a growing season of only around 60 days, hammered by 160kph (100mph) winds and receiving only 25cm (10 inches) of rain a year. It's hardly surprising that, after 1000 or so years, they can look ravaged. But this gives the clue to the bristlecone's survival tactics. It's been said that the bristlecone's strategy of long life is actually one of taking a long time to die.

Young trees less than a few hundred years old look very different. They have rich, reddish-brown bark and neat branches densely covered with glistening needles arranged in an elegant spiral that makes the branch tips look like foxtails – the name given to the group to which the bristlecone belongs.

The amount of sand-blasting and freezing that a bristlecone will experience over a 4000-year life will leave it with terrible scars and looking half dead. The oldest pines have few living branches and even less intact bark – it's not uncommon for a 12-metre (40-foot) tree to survive with only a ribbon of living

bark. Such a small amount of living tissue requires less food and water, and this, combined with needles that live for up to 30 years, means that the tree needs few resources to keep going.

It's hard to know what exactly brings the life of a bristlecone to an end. Really old trees have such tough,

resinous wood that they're virtually immune from attack by wood-boring insects or fungus. When they eventually die, they can stand for another thousand years without decay, their scoured skeletons ghostly-white. So tough are they that they erode like rocks. Its great age and almost indestructible nature make the bristlecone a supernatural natural wonder. The same characteristics also make it an invaluable natural archive of climatic history since the last ice age. Every year of its life the pine lays down a growth ring in its trunk, and the gap between each ring bears a very close correlation to the aridity and temperature of each summer it has lived through. Using living and dead bristlecones, it has been possible to track a continuous record over 10,000 years.

The fastest plant on the planet

Above *A forest of Japanese bamboo growing at such a rate that it's capable of reaching 30 metres in just a few months. The record for the fastest-growing plant on the planet is held by a species of bamboo.*

Opposite *Bamboo – just grass but with woody stems of extreme strength and flexibility.*

Is it possible to watch grass grow? It might be, just about, with one group – the fastest growing plant on the planet: bamboo. It's one of the commonest groups of grasses, with more than 1500 species – some tiny, some gigantic, some with individual woody stems reaching 23-25cm (9-10 inches) in diameter and well over 25m (82 feet) high. Sometimes it lives life at breakneck speed, sometimes it's extraordinarily cautious and patient.

Bamboos grow from underground rhizomes, either producing dense clumps or sending out runners 6m (20 feet) or so under the soil that, over time, form an interconnected forest of genetically identical individuals. Bamboos grow in an odd way. When new stems appear, they are not the normal delicate youngsters of most plants but emerge at what will be their maximum diameter. As they grow, the stems get no thicker. The tip is a series of tightly overlapping sheaths covering a predetermined number of leaf-bearing nodes that grow out telescopically like an extending radio aerial. What's more, the stem does all

its growing in its first season, and considering that might last only a couple of months, a 30m-high stem has to do a lot of growing. Super-fast growth may be an adaptation to take advantage of light gaps in the forest, but no one really knows.

But for all their speedy growth, bamboos take their time when it comes to sexual reproduction. Some species may only do it once in their lives and then only after more than a hundred years. Once a bamboo plant does get round to reproducing, it not only does it with abandon but it also does it at the same time as all its fellows – a habit known charmingly as gregarious flowering. It was once thought this involved all individuals of the same species in all populations, but it's now known that such mass flowering is limited to only local populations.

The significance of synchronous flowering is more about seeds than flowers. A bamboo may only produce seeds once, but it makes up for it in the massive amount it produces. A 33-square-metre (40-square-yard) plot of one species of bamboo was found to produce 136kg (300 pounds) of seeds – at least 4 million of them.

The reason for such prolific production may be to swamp seed predators – producing far more than the seed-eaters can consume and so guaranteeing that enough survive – though why bamboos should save up all their seed production for one single event is less clear. It may be that the effort to produce so many seeds makes such a huge demand that they can only manage it once. Certainly, the reproductive effort exhausts them so completely that they they die. But once a seed is set, a tropical bamboo can produce a full-grown stem in just 45 days.

So how fast can bamboo really grow? *Phyllostachys bambusoides* holds the record, with one plant reaching 1.2m (4 feet) in 24 hours, a speed that would be just about visible to the naked eye. So you really can watch grass grow… as long as it's bamboo.

The ancient dragon trick

Above *The dragon's-blood tree of Socotra, beautifully adapted to its hot and arid island home. The funnel-shaped arrangement of branches and gutter-like leaves collect water, channelling it down to the roots. If damaged, the bark oozes blood-red sap – the dragon's blood.*

Had Salvador Dali ever taken to biological illustration, the Socotra dragon's-blood tree would probably have been a favourite subject. The shape of its canopy (like an inside-out umbrella), the colour and shape of its leaves, its dramatic landscape setting and, of course, the blood-red resin that exudes from its bark give it an aura of the surreal. The only place where it exists is the equally extraordinary Socotra archipelago – the 'Galapagos' of the Arabian Sea. Lying off the coast of Yemen, the island of Socotra is a remnant of Africa cast adrift when Arabia and Africa split apart at least

10 million years ago. Having been isolated for so long, it has a large number of species found here and nowhere else. The landscape is brutal, prehistoric and parched by the equatorial sun. Any soil is thin, sandy and stony. Yet the crags and gullies are perfumed by frankincense and myrrh. Peppering the landscape are Socotra desert roses that appear to grow from the rock without roots or leaves, their pink flowers topping short, squat trunks swollen with water. Overshadowing them, in the mountainous region, is the Socotra dragon's-blood tree, up to 6m (20 feet) high and perfectly adapted to

this harsh environment. Considering the climate, it's hardly surprising that the tree is slow growing. In fact, it takes about 200 years to reach maturity.

Though the island is very arid, the mountains do receive occasional sea mists, and there is a brief twice-yearly monsoon drizzle. The dragon's-blood tree is shaped to make the most of every drop of that water. Like a giant funnel, it reaches up to catch the rain. Its spiky leaves are like gutters, angled up and so densely packed that they channel any water that condenses or

falls on them directly to the centre of the tree's crown, from where it flows down to the roots. They are fleshy, thick-skinned and covered with a waxy cuticle to both reduce water loss and speed the flow of water along their surfaces. The canopy is so dense that, once the rain has stopped and the full force of the sun returns, it acts as a parasol, shading its own roots.

It's an extraordinary plant and a perfect example of design moulded by a landscape and climate that have remained unchanged for millennia.

Above *The dense, parasol-like canopy of a dragon's-blood tree, shading its roots from the sun.*

Next page *The brutal landscape of Socotra. Strange desert roses, with trunks like expanding water bottles, grow alongside the dragon's-blood trees. Both plants are found only in this Arabian archipelago.*

Poisoning, strangling and twisting perversion

There are many solutions to winning the battle to gather light – growing faster than your rivals, having bigger leaves, strangling and even poisoning – but perhaps the most dynamic is to use your rivals' own bodies to get to the light above them.

Climbing is an economical strategy. By getting another plant to do all the work, you can put your energy into making leaves instead of building a strong supportive stem. Climbers use any and every part of their anatomy to get a purchase on their hosts. Ivy uses its roots, honeysuckle simply winds its stems rapidly round its competitors, others modify their leaves into long, often extremely mobile, tendrils, some with hooks, some with sticky pads. One of the most elegant of all climbers is the passion flower. As it searches for a victim to climb, the tendrils on its growing stem flail around, groping for any contact. These tendrils are both very sensitive to touch and also surprisingly selective. If a tendril grabs a structure that it senses is too smooth to provide a secure anchorage, it will

unwind itself and start a search for a better purchase. Once satisfied that the anchorage will hold firm, it rapidly wraps its tip round the host's stem and then begins to coil its mid-section first in one direction and then in the opposite direction – creating, in effect, a spring. This has two advantages. The spring acts as a shock-absorber, avoiding breakage of the fine anchor points, and it pulls the passion flower's own stem closer to its host's, so that other tendrils have an easier job catching hold.

The process of coiling in alternating directions has fascinated observers, including Charles Darwin, for years. Botanists call it free-coiling, but mathematicians call it 'perversion'. The process is remarkable because it creates a twist-less spring: by coiling one section clockwise and the other anticlockwise, the two sections cancel each other out, and so there is a net twist of zero. For a plant that wants to make a spring out of a tendril anchored at both ends, twisting perversion is the only way.

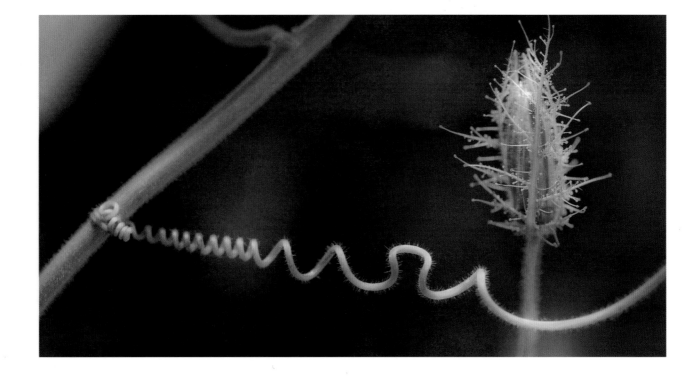

An aerodynamic marvel

Almost all plants are rooted to the ground, which makes colonizing new areas problematic. Though a plant may not be able to move itself, it can move by proxy – by spreading its seed. And within its seed, the plant has all the genetic information needed to reproduce and invade new territory.

There are almost as many ways of getting a seed to a new location as there are types of plants: hitching a ride, wind-sailing, water-rafting, parachuting, helicoptering, gliding and even ballistic catapulting. One of the most spectacular flying seeds is that of *Alsomitra metacarpa*, the Javanese flying cucumber, or climbing gourd, in the forests of Borneo. *Alsomitra* is a liana (vine) that climbs tree trunks to get to the light at the top of the canopy. It produces fruits almost as big as footballs, containing hundreds of wafer-thin seeds, stacked together like playing cards. When the fruit is ripe, it splits open, and with each breath of wind, a

handful of the seeds spills out. Each seed is a paper-thin glider with a 13cm (5-inch) wingspan (the seed itself is just a couple of centimetres wide). Once launched, the glider begins its flight by swooping some distance away before slowing into a lazy, spiralling descent.

The distance it travels depends on the height of its launch, the wind conditions and any obstacles, but surprisingly long journeys are possible. The great height of the canopy probably makes this type of gliding seed-dispersal effective. Sometimes, however, an *Alsomitra* seed adopts a different type of flight. During its glide, it suddenly goes into a stall and nosedives. As it does, it gathers enough speed to generate lift so that it soars upwards for a metre or so before stalling again and repeating the whole cycle. There is a lovely sense of anticipation watching *Alsomitra* fly in this rhythmic stall-dive-soar pattern. Not only does it seem as if the air is filled with transparent bird-wing butterflies, but the seeds also appear to fly further.

Gliding is actually quite rare in nature. Most flying seeds are either parachutes or autogyros. It's rare because it's hard to make a glider stable, which makes the *Alsomitra* seed all the more remarkable. Aerodynamic analysis has shown that it has what aircraft designers call a dihedral to its wings, so that the tips are raised above the body, which gives stability, and the wings are also swept up at the back on their trailing edge, which helps it recover from stalls, disturbance from turbulence or bumps into branches. This, combined with the incredibly light wings and resulting low glide angle, has led to its distinction as the most ingenious of all flying seeds.

So impressive is it that aviation pioneers took its design as inspiration for their early attempts at flight. Igo Etrich built a tailless glider in 1904, made from bamboo and canvas, in the shape of an *Alsomitra* seed. A modified version of his glider is credited with making the first true flight of a manned aircraft.

Wind-runners

Right *Brunsvigia in bloom.*
The mass emergence of these
fast-growing flower spikes from
their bulbs deep underground
is triggered by rainstorms.
The bright pink petals advertise
nectar to attract pollinating
moths at dusk.

African brunsvigias are the mayflies of the plant world, spending most of their life hidden away and only showing themselves briefly and spectacularly when it's time to breed. In a strange habitat known as succulent Karoo, in the arid Western Cape of South Africa, lives *Brunsvigia bosmaniae.* Like many of the plants in this area, it is finely tuned to survive the Karoo's challenging climate. Temperatures are high all year round and rainfall is poor and short-lived, falling mostly in brief winter and autumn storms.

In winter, all that can be seen of a brunsvigia is a quartet of fleshy leaves lying flat on the ground – like a giant green butterfly with its wings open. But below ground, a grapefruit-sized bulb is being fed by the leaves. This is the strategic food store that allows the plant to survive both the heat and droughts of summer.

As spring quickly turns to summer, temperatures soar, the soil bakes and the leaves wither. Everything now depends on autumn rain. The dormant plant needs a sharp storm to trigger it back to life. If there's a really good thunderstorm in mid-February, then the clock begins ticking, and almost exactly three weeks after the deluge, the soil is pierced by a flower spike. Hundreds of the spikes appear, apparently synchronized to emerge en masse.

The flowers grow extremely quickly, so fast that they appear to change shape in front of your eyes. The buds open up into huge globes of deep-pink tubular flowers – hundreds and hundreds of football-sized pink lollipops, creating an extraordinarily beautiful but strangely out-of-place riot of colour. The narrow tubular flowers are delicately lined with deeper pink. They don't appear to be attractive to bees, but as dusk falls, delicate noctuid moths arrive to sip the nectar, pollinating them in the process.

The hellish temperature means the flowers shrivel within a few weeks, but once fertilized, seeds start to form. The pink lollipops turn to desiccated pom-poms

Above *A brunsvigia flower globe transformed into a ready-to-roll seed disperser.*

Opposite *Ripe seeds. They will start germinating immediately they touch ground – making the most of any moisture left in the soil in advance of the baking summer.*

of seed capsules. You might imagine that the pea-sized seeds are simply shaken from the seedhead by the wind. But brunsvigia's strategy for spreading its seed is much cleverer than that. True, the wind plays a part, but instead of simply shaking the seedhead, it snaps off the entire globe, which then rolls away, scattering seeds as it goes, along with dozens of other pom-poms, spreading seeds over the Cape hillsides.

With such a short growing season, every day counts, and so the seeds themselves have one more adaptation to survive in this landscape. They are what horticulturalists call recalcitrant (they can't be stored), meaning they germinate the moment they come into contact with the ground. Just one month after the first appearance of the plant above ground, a new generation has begun the cycle all over again.

The numerate, seductive Venus

A fly's reactions are among the fastest of any animal – 20 milliseconds from the moment it registers movement to take-off. Yet it falls prey to a hunter that can move even faster. Plant movement is generally too slow to be seen, but with the Venus flytrap, it's simply too fast.

Like most carnivorous plants, the Venus flytrap is adapted to survive waterlogged and acidic places where it's hard to get nitrogen (essential for building tissues) from the soil. In a neat reversal of the usual food chain, it has become a hunter, obtaining nitrogen from the bodies of animals. Its trap is a leaf that, as it grows, undergoes an amazing metamorphosis. First it swells as if pumped up, then it splits along one edge, opening like a clam. The two edges then start to develop spikes like green eyelashes, and the inner surfaces grow stiff, thin hairs. The hairs are the triggers that will spring the trap. The final addition is bait: along the edge of the leaf grow tiny nectaries that secrete a sugary fluid irresistible to flies. The trap is set, and the plant waits.

A foraging fly senses the nectar and approaches. To get to the nectaries, it passes between the jaws. As it does so, it brushes against a hair. But the trap doesn't close. The fly pauses, perhaps to clean its mouthparts, then moves along a little, clipping two hairs in quick succession as it does so. Instantly, the trap jaws snap shut, the 'eyelashes' meshing firmly together. There is no way out for the fly, and as it struggles, the trap shuts

tighter, sealing to form a kind of leafy stomach. Then the plant begins to feed. It secretes enzymes that digest the fly. Body fluids are absorbed by the leaf until all that's left is the empty husk of the fly. The plant now resets its trap, cranking open the jaws by pumping fluid into the cells, and in a final macabre act, it 'spits' out the corpse.

Learning to count is a way to avoid false alarms. Requiring two hairs to be touched in quick succession increases the chances of the trap being triggered by a suitably large insect and not a midge or, worse, an inanimate object. Should a tiny insect trigger two hairs, the plant has another size filter. When the trap shuts, there's a pause before it's sealed, allowing small insects to crawl out through the grill of interconnecting spikes. If the prey leaves, there is no further stimulation of the trigger hairs, and the trap reopens and resets.

This is an effective strategy for obtaining nitrogen, but it does have one drawback. When the time comes for the Venus flytrap to flower, it needs insects to carry its pollen to other flowers. So how can it avoid eating the very creatures it now wants as allies? The answer becomes clear when it starts to produce its flowers. The plant sends them up on a long stalk, long enough to hold the flowers well clear of its deadly jaws. Pollinating insects can then fly in high enough to feed safely and pollinate the flower without being tempted by the traps below.

Opposite *A deadly trap, primed and on a hair-trigger. The fly, attracted by nectar extruded around the edges, has already touched at least one of the six or so sensitive trigger hairs on the inside of the leaf. It is doomed.*

Below *The trap closed tight – and opening up again. When the fly touches two trigger hairs in quick succession, a chemical-electrical signal causes the trap to shut instantly. Entombed, the prey is slowly dissolved in a soup of enzymes. Once the plant has finished feeding, the trap opens, and the insect husk blows away.*

Next page *A last, desperate attempt at escape. To make sure the prey is big enough to be worth catching, the Venus flytrap delays the final closing of its jaws, leaving a gap so a tiny insect can escape between the bars. This big fly, though, is destined for dinner.*

chapter 4

Insect ingenuity

INSECTS RULE THE PLANET – a view held by many who study invertebrates. They claim that the idea of a present-day 'age of mammals' is little short of scandalous and that there never was an 'age of reptiles'. They say the age of insects began 400 million years ago, long before mammals and reptiles existed, and it has never ended. It's not an argument that can be settled, but if you play the numbers game, the phenomenal diversity and abundance of insects, and the scale of their impact on natural systems, cannot be denied. Around a million kinds have so far received names. Speculation about the total ranges from 4 million to 40 million. Even the lowest estimate gives you 888 times as many kinds of insects as there are mammals. There are around 200 million insects for every person alive and 3 billion per square kilometre (a third of a square mile) of habitable land. The ecological services they perform are almost unquantifiable. Take honeybees. The workers of a single colony may make millions of visits to flowers every season. Humans harvest the produce of only a few species of bee-pollinated plants, and yet the value of these is about $50 billion a year.

The insects' secret is their ability to produce a diversity of body-form and behaviour unmatched in the animal kingdom. There are a number of reasons for such flexibility, but first among them is their skeleton – not a hard frame inside upon which the working parts hang but an outside shell around them. The major component of this exoskeleton is chitin – a polymer, like plastic, with the same mouldability, which can be elastic or as hard and rigid as some metals. It can be reshaped with little effect on the functioning of the soft parts. This allows insects to create tools out of parts of their bodies. The massive weapons of stag beetles are enlarged and reshaped jaws. A praying mantid's ancestors would have had normal front legs, whereas present-day ones have muscle-filled weapons like gin-traps. Insect wings are outfoldings of the exoskeleton. Each compound eye of a whirligig beetle, which lives on the water surface, is split in two – adapting it for seeing both below water and in air.

One of the greatest advantages of this ability to reshape has been the development of a life-cycle of four entirely different stages, resembling different creatures. At each stage, the body is moulded to do only what it needs to do. No material, no behaviour, is superfluous. This is far more efficient than having just one body capable of doing everything. For example, a butterfly caterpillar just eats. It doesn't need the wings or fancy sensory systems of an adult. It's a food-processor that can increase its weight by 10,000 times in 3 weeks, rapidly storing up enough material to make an adult insect.

Another major factor in insects' success is their mastery of chemistry. All kinds of insects manufacture chemicals for self-defence, developing nozzles, jets and hollow hairs to deliver them. Insects also use chemicals in the form of pheromones to communicate. A female moth may have special extendible organs to disseminate sexual pheromones, and their males have complex antennae, like television aerials, which pick up the scent several kilometres away.

Left A gathering army of South African brown locusts – here at the hopper stage. When conditions are right and millions hatch, they release a pheromone chemical that transforms them from solitary individuals into social swarmers. Within a week, they will have developed wings and become a plague.

Previous page A male Darwin's stag beetle. His giant weapons, moulded from his upper jaws, or mandibles, are designed to lever rivals up and off their branches.

Below *A South American large-headed ant holding the head of an insect. It has excellent vision and stalks fast-moving prey, snatching it with powerful trap-jaws operated by a catapult mechanism. It can move its head from side to side in a way eerily reminiscent of a mammal.*

Right *The full display of a peacock katydid from Guyana, revealing another use for wings. It normally resembles a dead leaf, but if a predator tests the camouflage, it flashes its huge fake eyes – enough to scare off an inexperienced bird or lizard.*

Pheromones underlie the complex organisation of social insects whose colonies may number millions of individuals. They direct the gathering of food and the construction of complex nests, and they rally the colony for a mass attack on predators. But insects have one major limitation. The exoskeleton only works at a small scale. A human-sized insect would need an exoskeleton so thick that there would be no room for organs inside. Yet insects have made smallness an advantage. Individuals consume very little, and so where lots of resources are available, a species can build up vast populations. An ant nest may house several million inhabitants. A locust swarm may contain 50 billion. A pair of flies could potentially give rise to an 8km-wide (3-mile) ball of flies in just two years. Such a horrifying scenario does not occur partly because fly numbers are kept down by predators, chief among them being other insects.

The irony of insects is that it is probably their very smallness that prevents them being accepted as masters of the world. Yet the reason behind it, the exoskeleton, is what allows them to be, arguably, just that.

Right *A tiny parasitic wasp using her enormous back legs to preen her ovipositor (egg-laying 'needle'), her abdomen thrown back against her wings. She is a mere 1.5mm long and has just hatched out of the egg of a preying mantis in Cambodia, and will go on to inject her own eggs into those of another mantis, where they will develop into more miniature wasps.*

The big day of the damselfly

Insects invented flight on this planet 330 million years ago. They had already dominated the land for 70 million years. Now they could colonise the air, too. This opened up a whole new branch of evolution, as bodies and behaviour changed to populate a new environment. Other animals that subsequently developed wings had to sacrifice a pair of limbs to do so, but insects make theirs in a much more efficient way. They are created

from folds in the exoskeleton, which can be moulded into all kinds of shapes.

The air rapidly became an ecosystem every bit as complex as the land, and insects' wings became multipurpose toolkits. They allow an insect to move quickly from place to place, they're a means of escaping predators and a means of becoming predators, and they can be coloured so insects can signal to each other.

Damselflies, along with their close relatives the dragonflies, are ancient insects, scarcely changed over millions of years. They're a perfect example of how

Above Copper demoiselles on the lookout for flies and midges. This may be the only day of their adult lives when it's warm enough to hunt, but they have already spent two years in the stream as larvae – the feeding and growing stage of their lives.

Opposite The culmination of the adult existence: breeding. A male holds a female while using the tip of his abdomen to take a packet of sperm from his genital opening and place it in a pocket on his abdomen, from where the female can collect it.

winged insects take advantage of land and sky to live three-dimensional lifestyles. A damselfly may spend two years as an underwater larva, feeding and growing in order to become a winged adult that lives for only a few days. In a poor summer, the adult may experience just a single day warm enough to try to reproduce and continue the species. It is a day filled with challenges and danger.

The copper demoiselle lives around streams in southern Europe. It has been intensively studied by scientists and its daily life laid bare. This begins when the rising sun burns off dew that has collected on roosting damselflies and they start to heat up. This is a dangerous time because small birds hop among the grass and reeds picking off the inactive insects. Once warmed to flight temperature, damselflies perch, often many together, on twigs or blades of grass overlooking their stream. From here they spot the tiny flies and midges they feed on – their eyes register movement six times quicker than ours do.

When a damselfly takes off, its rate of acceleration and precision of aerial movement are among the greatest in the animal kingdom. It brings six bristle-lined legs together to form a basket that it uses to sweep prey from the air. Some damselflies might fall prey to spiders that spin webs in their flight paths. But the ones that feed and survive can then begin the main business of the day.

Males set up territories to attract females. A perfect spot is a small plant surrounded by underwater weeds. A male perches on the plant and attracts females towards him with his copper-green wings, hoping to impress her with the quality of the weeds on which she can lay her eggs. But since good territories are scarce, he also attracts other males. The defender takes off and angles his wings at them so they act as flags, flashing colour to warn them off. If this doesn't work, they grapple in mid-air. Each tries to push the other down into water, and drowning is common.

Below *A near-catch of a demoiselle. Frogs are among the adult demoiselles' chief predators, using their projectile tongues to catch them when they are occupied mating or egg-laying.*

Opposite *A female copper demoiselle. She is larger and heavier than a male, and her abdomen holds hundreds of eggs. At the end of the abdomen are 'teeth' for cutting holes in stems in which to lay her eggs.*

When a defending male spots a female approaching, he behaves quite differently. He flies around her, beating his wings 50 times a second, three times faster than normal. This allows her to assess his strength. Damselflies only breed in running water, and so he lands on it and floats past her to demonstrate the perfect speed of this bit of the stream. If she accepts him, she flutters her wings and they fly to the protection of dense vegetation to mate.

He seizes the back of her head with claspers on the end of his abdomen. Then he bends his abdomen in a loop until the tip meets the base, and he transfers the sperm packet there. The female bends the tip of her abdomen around to pick it up, their bodies now forming a heart shape. But before she can take up the sperm, the male scrapes at her to remove any older sperm from previous matings. Marauding males may attack them and try to prise the male away. They scrape and bite, ripping off bits of his wings or whole legs.

Yet an even greater danger lurks under water, because frogs specialize in catching courting damselflies or egg-laying females. They launch from the water, shooting out a forked tongue to pluck the damselflies from the air.

If a pair can survive birds, spiders, frogs and other damselflies, then egg-laying begins. The female lands on a stem and backs down into the water. The male sits nearby to defend her. She submerges entirely, turning silver from the thin layer of air trapped around her, cuts into plant stems with her ovipositor and lays her eggs.

Once this is done she lets go and floats to the surface. Her wings may get stuck in the surface tension as she tries to take off, leaving her vulnerable to attack from below by water beetles and water boatmen. Whether or not she escapes to live another action-packed day, she has survived long enough to leave offspring that will eventually give rise to more short-lived damselflies.

The chemical warrior and the screaming mouse

Among the more remarkable invertebrates are the ones that have become walking chemical weapons. Their exoskeletons are moulded to create spines, hollow hairs, stings and rotating sprayers, all for delivering a range of thoroughly unpleasant substances. Sometimes this is for defence, sometimes for overpowering prey.

The range of chemical weaponry is stunning. Ants spray formic acid; the bombardier beetle has an internal chemical factory every bit as wondrous as its pinpoint aim of toxic, boiling spray; the spines of the lonomia caterpillar have an anticoagulant so powerful it can kill people; stick insects spray terpenes; spiders spit venom. The list goes on and on. But for all-round excellence, few can match scorpions – not insects, though closely related and with exoskeletons. All use deadly venoms made of neurotoxins to subdue prey, which affect the nervous system and lead to paralysis. Some don't affect humans, while others, such as that of the death stalker, are lethal. The bark scorpion was responsible for about 1000 human fatalities a year in the southern US and Mexico before an antivenom was created.

Despite poor vision, the scorpion has a bank of senses to locate accurately the direction and distance of prey or of danger. Two organs touch the ground and track scent trails. Fine body hairs and slits on the scorpion's legs sense minute vibrations through the soil and help it work out the distance to its source. Specialized hairs on its front pedipalps detect airborne movements and give accurate information on its direction. A scorpion's tail-sting whips in any direction to jab at predator or prey. Traces of the metals zinc and manganese from its food laid down in the tip of the sting allow it to punch through tough skin or cuticle and deliver the venom.

Scorpions might appear almost invincible, but on dark nights in the southwestern US, they encounter their ultimate adversary, one that is neither poisonous nor armour-plated and weighs only 14 grams (half an ounce). The grasshopper mouse is the fiercest and quickest predator in the desert, and its battles with the desert hairy scorpion, the largest of the region, are as dramatic as anything on the African plains.

The grasshopper mouse has some very unmouse-like traits. It eats little else but meat – from grasshoppers and beetles to ants and scorpions. It needs to defend a territory large enough to live off and can't afford to be shy. Instead it stands on its hindlegs and screams. The sound carries 200m (656 feet), far enough to reach the most distant boundary of its territory. This warns off intruders as well as attracting mates.

When hunting, the mouse has the same strategy as larger predators. It stalks prey, then rushes in and immobilizes it with a bite to the head. Most victims rapidly succumb. But not the desert hairy scorpion.

Right *A grasshopper mouse stalks a hairy desert scorpion. Its aim is to bite off the scorpion's venom-filled sting and then kill it. But the scorpion is wise to that.*

Left *A meat-eating grasshopper mouse, screaming a warning to rivals. It's a predator that even armoured and armed scorpions have reason to fear.*

It faces the oncoming mouse, sting raised high, then lashes out. The mouse ducks, weaves and sways like a boxer. A scorpion's jab may appear certain to strike only for the sting to slide harmlessly past the mouse's face or body, as the predator takes evasive action.

Occasionally a scorpion will land a sting, but the mouse just carries on. The venom may cause pain, but it isn't lethal, as the mouse is resistant to the venoms of the scorpion species in its region. Probably this resistance is partly inherited from its mother and then reinforced through exposure to venom during fights. This puts the scorpion at a distinct disadvantage since it can't rely on its primary weapon to end a fight. All it can do is try to keep the mouse at bay. Also, the mouse has a scorpion-specific attack strategy. Instead of going directly for the head, it tries to grab the base of the sting. But it's an evenly matched battle. If the scorpion tires, the mouse will take its opportunity and bite off the sting, but if the scorpion is strong and healthy, it may fight long enough to drive the mouse away.

Scorpions have been around for at least 300 million years and have scarcely changed their form, proof that their stings are a very effective means of defence, even against exceptional killers such as grasshopper mice.

Left *The scorpion whips round to sting its attacker. The mouse is resistant to the venom, but the counterattack keeps it at bay.*

A working mother's ultimate sacrifice

Above *Japanese parent bugs clustering, becoming one big red and black message to predators: 'bad-tasting food'. Females will cease being cooperative when raising young and will steal food from each other. But they may also make the ultimate social sacrifice – providing their bodies as food for their young.*

A major factor in the success of mammals and birds is the time and energy they spend caring for their young. Insects usually take a more basic 'lay them and leave them' approach. They rely on sheer numbers – out of hundreds or thousands of eggs, a few will make it to adulthood. But in certain difficult conditions, this could be disastrous. It's a mark of insects' flexibility of behaviour that they deal with the problem by adopting parental care. Usually this is just for a short time, but a few species take it to almost mammalian levels.

One of them lives in the forests of Japan's south island, Kyushu. It's a brilliant red and black shield bug. First described in 1880, it remained just a bug for 100 years. Then a life history of startling sophistication came to light. Its existence depends on an unreliable food source, the fleshy fruit, or drupe, of just one kind of tree of the family Olacaceae. The problems this causes the bug have driven the complexities of its lifestyle.

The bugs feed on fallen drupes. When the drupes fall depends on the weather, and only as few as 5 per cent will be in the right condition to be eaten. As if this weren't enough, the tree provides another challenge. A female times her egg-laying so her babies hatch just as the drupes they must eat are falling. Her hiding spot for them is a natural hollow in the leaf-litter. But the Olacaceae tree drops its leaves at a different time, and so the area where the fruits fall is bare. As a result, a mother bug must find a nest spot elsewhere, up to 12m (40 feet) away. If she were a normal insect and her tiny offspring had to travel that far to find food, exposed in the open, most would be picked off by predators. The mother has no option but to find their food for them.

The female bug stays with her eggs and protects them from predators, chief among which is a ground beetle. If a beetle approaches her nest, she makes a scraping sound by dragging her wings against her body and turns her back to form a barrier. If the threat continues, she runs away carrying her eggs. This protective behaviour is not so unusual – other bugs in her group

do this, too. But once the eggs hatch, things get more interesting. The mother must find drupes. After journeying to the tree, she might spend hours probing fruits before finding one she likes. Jabbing her mouthparts into it, she then drags it away. This is a formidable task, since the drupe can weigh three times as much as she does. Unfortunately, though, her problems have just begun. Since suitable drupes are rare, it pays a female to ambush another bug that has one, rather than wasting time looking for one herself. A female dragging her drupe can be set upon by half a dozen others, leading to a multiple tug-of-war. The winner still has the problem of finding her way back to her nest. Yet however meandering her outward journey, she always returns in a straight line. She does this by memorizing visual markers in the canopy above and uses them as a map to find the shortest route.

The young bugs in the nest swarm over the new drupe, but it doesn't last them long. Their mother must constantly supply new ones as they grow. A successful nest can contain the remains of 150 drupes, each hard-won. Not every female, though, is such a model toiler. Some are nest thieves. In years when good fruits are rare, these females resort to raiding drupes from nests unprotected by their hard-working owners.

If a female can overcome all these challenges, she will supply her young with enough food for them to reach independence. But the young themselves have a say. If they judge that their mother isn't bringing back enough good-quality drupes, they simply leave. They search for a nest supplied by a more successful female and join it. The new mother accepts them even though they have just doubled her workload. This might seem to relieve their real mother of her relentless searching, but she is so hardwired to find food that she just carries on, piling up uneaten drupes in her empty nest. But the adopted mother comes out of it even worse. Having managed to feed so many young to the point of independence, they turn on her. A mother, or foster mother, is her offsprings' last meal before they leave home.

Left *A mother's labour of love. Every day is a relentless search for drupes at the right stage of ripeness for her babies to feast on. A drupe may be several metres from her leaf-litter nest and nearly three times her weight. She must drag each one back, defending it from thieving females on the way.*

Bee life in the Outback – brawlers and chancers

The physical toughness and adaptable lifestyles of many insects allow them to thrive in the harshest places, but it can mean adopting some unusual and sometimes deadly strategies.

The western side of Australia is a remote and tough landscape by any standards. Halfway up the coast lies a huge region called the Kennedy Ranges. The sandstone rock there contains fossils that reveal its origins as a shallow sea basin. Over time it rose until the western side became a huge plateau, now eroded into gorges and vertiginous rock faces. The eastern part of the Ranges is a forbidding, arid plain. Further still from civilisation, it is remote even by Western Australian standards. But what this area lacks in gaudy geography it makes up for in fine detail.

Its most striking features are red-clay pans. They form when rare rains create low-lying pools. The water evaporates, leaving billiard-table-smooth expanses of clay. The footprints of a wandering kangaroo or emu may sometimes be impressed in their surfaces before

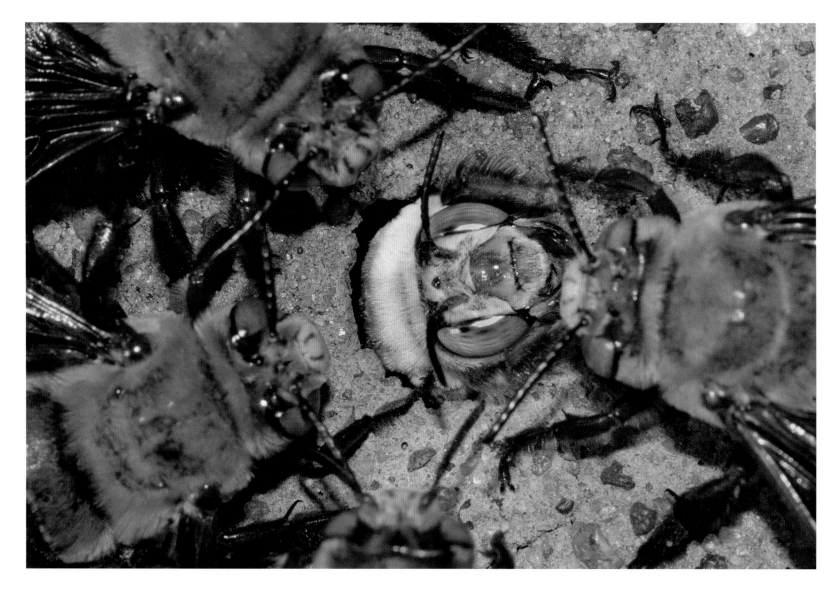

Above *An emerging female (with the white head) surrounded by males waiting to grab her. Males emerge first and smell out females as they dig their way through the hard clay.*

the clay hardens like concrete. Some pans are circular, others indented like the outlines of countries. With no water, no vegetation and no hope of shelter from the sun, the pans are just about the most inhospitable places in the whole Kennedy Ranges.

Yet for a short period in the hottest part of the year, hundreds or thousands of small pyramids erupt all over a few of the pans. A closer look reveals numerous giant white bees zooming around. Every so often one

will hover over a pyramid, then drop out of sight, giving the place the impression of a busy helicopter airport. The pyramids are the spoil heaps from underground tunnels dug by Dawson's bees, one of the largest and most beautiful kinds of bees in Australia.

The precarious nature of living in such a place has demanded extreme strategies. The bees have found a clever way to dig into the baked earth. They gather nectar from some of the few flowers found here –

such as hakea, eremophila and northern bluebells – which they use to soften the clay so they can bite out jawfuls. Over several days, a female digs down and then sideways, forming a branching network of tunnels. Then she collects nectar and pollen, which she stores at the bulbous dead-ends of the side tunnels. An egg is laid in the sludgy mass at the end of each brood cell. When the cells are sealed off, her life's work is done. The egg produces a grub that grows on the stored food and, next season, hatches out as a bee.

Dawson's bees have a near-infallible ability to find their own pyramids, probably by memorizing the miniature landscape of their pan. But they do occasionally get it wrong, leading to mid-air collisions or brief ownership battles when a bee goes down the wrong hole. They are not social like honeybees. The pans resemble commuter estates where the occupants are always too busy to make friends with the neighbours. But this harmonious metropolis is founded on an uncanny fact. Every bee is female. The reason for this lies in a violent past.

Above *A violent brawl over a female. Competition among males is so intense that, in the heat of battle, an emerging female may get caught up and accidentally killed.*

The males emerge a month or two earlier than the females, when the clay pans are still unblemished by tunnelling. Some are large and some are small. They fuel up on nectar, and then the males of different sizes pursue different strategies. The larger ones patrol the pans and the smaller ones disappear.

Eventually the females begin to hatch underground and start tunnelling to the surface. The moment a female makes the first tiny hole at the surface, a nearby male can smell her. He lands beside the opening and waits for her to chew her way out. If he is very lucky, the other males will be busy elsewhere. But usually they quickly appear. Competition for females is intense, and the first male tries to prevent the newcomers from landing by flying backwards into them. But if he is besieged and they start to land, he has no option but to fight.

It is rare in the animal kingdom for individuals of the same species to kill each other in combat, but Dawson's bees don't obey this rule. Males grapple with

each other, rolling over and over across the clay. Their weapons are a sting and powerful jaws. A male may be driven off by the ferocity of an attack or he may be badly injured or killed. If males are abundant, brawls break out involving up to a dozen bees in a tumbling ball attacking each other indiscriminately. Each male is trying to get himself in pole position beside a burrow for the moment when a female finally breaks out. Then he'll drag her away to the edge of the pan, where they can mate under the cover of shrubs. But if the violence gets out of control, the female will be dragged into it, and is as likely to be killed as a male.

Yet not all males take the violent route. The smaller ones are unsuited to fighting and have a sneakier strategy. They lurk at the edges of the pans, well away from the fighting, waiting for luck to work their way. Occasionally the big males are so busy fighting that a female emerges and escapes unmolested. Then she'll be grabbed and mated by one of these satellite males.

Life in this environment is so difficult that it is essential for every female to be mated to maximise the number of offspring. Producing different kinds of males is the most effective way of ensuring this. The males' size is controlled by their mothers the previous year. Early in the season, when food is abundant, a female digs larger bulbs at the ends of her side tunnels and stocks them with large amounts of food. These produce the bigger males. But as the food supply diminishes later in the season, the female digs smaller bulbs and stores less food. The result is small males.

Deserts are fickle places, and despite the bees' strategies for survival, their existence is precarious. A local drought might kill the flowers around the pans where the bees hatch, or a grasshopper plague might devour every plant. Then the females have to disperse across the desert, hoping to find a pan where flowers have survived. The bee population will crash until better conditions allow them to build up their numbers again.

The Argentinian antropolis

Above *The march of* Atta, *bearing grass for the metropolis. In the 'kitchen' chambers, fungus turns the grass into food that feeds the 7 million or so ants.*

Aerial photographs of the palm-speckled grasslands of northern Argentina reveal large white discs scattered randomly. Radiating from each disc, and connecting some, are what look like trails. It's an image reminiscent of satellite photos of cities at night: glaring with light pollution and connected by the white streaks of busy roads. The comparison is valid, because each of these white discs may be home to 7 million inhabitants, who harvest the produce of the land and transport it along the maintained roadways to fuel their metropolis. These are the grass-cutting ants. Their mounds may be 5m (16.5 feet) wide, white with pale excavated soil.

Such sociality is the peak of insect evolution – and the closest an animal comes to matching the complexity and scale of a city. Bees, wasps, termites and, of course, ants all do it. That they have been able to achieve this feat is due to a number of reasons, among them the flexibility of their exoskeletons and their special chemicals.

Vast colonies demand a constant flow of fuel – grass, in the case of the Argentinian ants. This is where the flexible exoskeleton comes in. It allows the species to exist in several forms, each designed for a different job.

These include the big queen, workers and a super-sized version of the workers with huge heads and jaws. The queen stays in the colony laying eggs, but every day in the wet season, and at night in the dry, huge numbers of workers march down their trails into the surrounding grassland. The large-jawed ones climb grass stems and cut through them. The severed sections topple to the ground, and the smaller ants collect them. They march down the trail holding the stems vertically, resembling phalanxes of Roman soldiers carrying their lances. The first ants to pick up the pieces don't carry them all the way back to the nest. Instead they take them part of the way and drop them, to be picked up by the next team in the relay. This appears inefficient, since a piece takes longer to reach the nest than if one ant carried it all the way. But the actual rate of transportation may be greater this way, and it increases opportunities for communication between ants. A big colony harvests 500 kilograms of grass fragments a year, making these ants the dominant grazers of this landscape.

Coordinating so many individuals efficiently seems an impossible task without speech. The key is the insect's ability to manufacture chemicals. When smoke from a dry-season fire engulfs a trail of marching ants, the means of coordinating their harvesting is revealed. The ants immediately drop the grass and, rather than racing back to the safety of the nest, they mill about aimlessly. The smoke has interfered with their pheromones – the chemicals they use to communicate. Each ant lays down a pheromone trail as it runs, telling other ants the direction of new grass. A great amount of pheromone signals that many ants have gone that way, so the grass must be of good quality. When the pheromones are disrupted by smoke, the wondrous order of millions of ants is sent haywire. But it's such a simple system that, once the smoke is gone, the whole thing is soon up and running again.

Unable to digest grass, the ants carry it deep under ground to special chambers. Here the big-jawed workers slice it up into little bits and stick it into

fuzzy white fungus-garden balls. The fungus breaks down the grass, growing on the nutrients this releases. An antibiotic in the ants' saliva prevents any other kind of fungus growing. The queen lays her eggs in the fungus garden, and her young feed on the fungus, which also provides a large part of the adult ants' diet. Neither ants nor fungus can exist without each other. But as well as giving the colony life, the fungus is also one of its greatest threats. In breaking down grass, the fungus gives off a lot of carbon dioxide, poisonous to the ants at high concentrations. To combat this, the ants build a ventilation system: vertical tubes on top of the nest that connect the interior to the open air. The tallest ones in the centre are designed so that, when

the wind blows across their entrances, air rich in carbon dioxide is sucked out. Around the edge of the nest are smaller tubes. Fresh air is drawn down them to the inside, keeping the atmosphere healthy.

The huge nest is like a fortress. Even the giant anteater doesn't try to get in. But the ants on the trails attract a different kind of killer. Minute phorid flies patrol the trails like fighter planes. Five different kinds mount aerial attacks. One will swoop down and, in less than a second, lay an egg on an ant. The bewildered target, its jaws flung open, stands still for several seconds, but by then the time bomb is planted. The egg will hatch into a grub that bores into the ant and eats it from the inside.

Above *Trunk-trails radiating out from the enormous colonies, excavated by the coordinated action of millions of ants – a testament to the capabilities of social insects. Ants are likely to be the dominant grazers of this grassland landscape.*

The world's biggest flutterby

Above *Overwintering adult monarch butterflies, warmed by a burst of winter sun, drinking from a pool of water in the forest. Some may drown in the rush to drink.*

Opposite *A daily flight on a warm spring day in Michoacan, Mexico. A billion butterflies may gather here.*

The monarch butterfly's great migration across North America is one of the wonders of nature. The distances seem too enormous for such a creature, yet the migration's existence is the perfect demonstration of insects' capacity for extreme behaviours.

The monarch's life-story reverberates with such contrariness. Its brilliant orange wings with their black veins and white-spotted black borders have been compared to stained-glass windows. Birds, on the other hand, find them disgusting. The purpose of the ornate pattern is to warn that the butterfly is packed with toxic cardenolides – unpleasant chemicals that will make a predator vomit. The human story surrounding the monarch is similarly contrary.

For several hundred years, North American observers have noticed how the butterflies become restless in autumn, gathering in unexpected places for a few days, sometimes in vast numbers, to feast on flowers.

Then they are gone and do not reappear until spring. In 1885, naturalist John Hamilton came across an especially large autumnal pitstop at Brigandine, New Jersey. He estimated the butterflies formed a band 4km (2.5 miles) long and 366m (400 yards) wide. Western science kept asking the question: where do they go?

It was only in 1975 that experts discovered the now-famous hibernation sites in Mexico. But amid the rejoicing and astonishment came a realisation that it wasn't entirely Western science's question in the first place. At the other end of the migration, the Purepecha Indians of Mexico's Michoacan state had been asking themselves the same thing for much, much longer.

Every October and November, they watched parts of their silent montane forests fill with hundreds of millions, perhaps even a billion, monarch butterflies – their collective wingbeats like leaves in a storm. The Purepecha called them the returning souls of their dead. But come February, the butterflies disappeared. They wanted to know where they went.

So when American naturalist Kenneth Brugger climbed above the 3050m (10,000-foot) contour of the Michoacan mountains in February 1975 and became the first Westerner to see the world's greatest congregation of butterflies, he could give two answers to the same question. He could tell the Purepecha that their returning souls flew north and that they gave rise to several more generations of northward-flying butterflies.

Ultimately the great-great-grandchildren of those Mexican monarchs reached as far as southern Canada, 4830km (3000 miles) away. Western science learned where their southbound monarchs went after they disappeared somewhere south of the Rio Grande in Texas. Scientists proclaimed the hibernation sites the eighth wonder of the natural world. But this was news to the Purepecha. Since they never travelled out of their area, they assumed there were places like it everywhere.

Monarchs are found in many parts of the world on either side of the tropics, and it would be easy to assume that every population makes massive migrations. But only one does – that of the North American continent. Why? The answer lies in the butterfly's past. Originally the monarch and the milkweed plants its caterpillars eat were found only in the hot regions of Central and South America. But 24 million years ago, the milkweed spread up into North America, acquiring frost tolerance along the way. The butterfly followed the food bonanza but never developed the frost tolerance. So on some unknown cue each autumn, the continent's entire monarch population migrates south ahead of the approaching winter. Guided by a combination of sunlight and magnetism, the majority find their way to the same groups of trees in Mexico where their ancestors have gone for thousands or millions of years.

Fred Urquhart, the mastermind behind the discovery of the hibernation sites, marvelled when he first saw them that 'such a fragile, wind-tossed scrap of life could have found its way across prairies, deserts, mountain valleys, even cities, to this remote pinpoint on the map.' After all, it wasn't so long ago that the idea of butterflies migrating at all was considered absurd. English naturalists attributed the sporadic appearance of certain butterflies in their islands to the idea that they spent an indeterminate number of years as eggs. But enough observations of butterflies on the move were eventually made to convince experts that butterfly migration is a common event.

Proof that the idea was totally accepted can be found in the history of World War Two, when a great, yellow cloud was spotted rolling over the English Channel towards Kent. With memories of toxic mustard gas still fresh from the previous war, the government was kept updated on developments. The cloud rolled in, and there was widespread relief when it turned out to be nothing more than clouded yellow butterflies making another of their well-known northwards migrations.

Mountain forests don't seem a sensible place for a tropical butterfly to go when trying to escape the cold. But monarchs come to Mexico to shut down for the winter, and they move north when spring brings a fresh growth of milkweed. They need low temperatures to hibernate, but not too low. Their chosen spots, at a precise altitude just above 3050m (10,000 feet), are perfect. The tree canopy acts as both blanket and umbrella. It keeps temperatures low within the forest when the sun shines and yet retains enough warmth during freezing nights. Butterflies exposed to rain are susceptible to freezing, but the canopy keeps them dry.

These hibernation sites are an almost perfect solution to the butterflies' problems with freezing weather, but occasionally things go badly wrong. In a major snowfall, monarchs are knocked out of the trees, or whole branches packed with hundreds or thousands of them snap. On the forest floor, the butterflies get wet and freeze. A 12-day storm in January 2002 killed around a quarter of a billion of them, leaving a carpet of bodies nearly a metre (3 feet) deep. Fortunately the species is resilient, and a good breeding season boosted their numbers back to secure levels.

The greatest threat is a recent and familiar one. Deforestation pokes holes in the blanket-umbrella of the canopy, leaving the butterflies vulnerable to rain and freezing air. The hibernation sites are officially protected by a presidential decree, but enforcement is difficult. As with so many conservation issues, the situation is complicated by the basic needs of local people and their sense of dispossession when long-held lands are taken from their control. The situation is approaching a critical point, and if a solution isn't found soon, the hibernation spectacle may be no more. If the sites can be safeguarded, then many more people will have the thrill of witnessing the extraordinary sight that greeted Kenneth Brugger back in 1975. 'So many butterflies the forest is more orange than green,' wrote one expert. Yet even this simple picture contains an element of the contrary: poor Brugger was colour-blind.

Above *Oyamel firs coloured orange by butterflies. The forest at 3050 metres provides a winter hibernation site where the temperature is just warm enough at night and not too warm in the day. The canopy keeps off the rain and helps prevent the butterflies freezing. It's such a perfect refuge that some butterflies fly 4830km to get here.*

Opposite *Huge numbers of butterflies gathering to warm up in a sunny spot in the forest.*

Next page *The colony coming to life on a warm day. The biggest natural hazard the butterflies face is extreme cold and damp brought by the arrival of a cold front. But logging of their forest refuge is the greatest danger.*

Eye-spy stalk-talk

Size matters in the animal kingdom, particularly when it comes to male stalk-eyed flies. Members of this group of flies emerge from their pupae with small, compressed, soft eye stalks. But within minutes, they start forcing air into the stalks, which get longer and longer – the span often exceeding their body length – until, after about 30 minutes, the cuticle hardens, fixing the stalks solid.

The flies live in Asian rainforests, and during the day, they search for food such as yeasts and bacteria on the surface of decaying vegetation. Having widely spaced eyes on the end of stalks would seem to offer no great advantage for finding such food. It's in the evening, though, that their eyes really come into their own.

As the light fades, stalk-eyed flies fly to nocturnal roosting sites – the exposed, threadlike rootlets that hang down under the eroded banks of rainforest streams – using their exceptional eyesight to avoid spider webs and other hazards in the forest. Night after night, individuals return to the same roosting threads, and some do so for many months.

Each evening, the males fight for control of harems of females. The largest males arrive first, and as dusk advances, the females and smaller males appear. Females fly between different males, before settling with their preferred suitors. Their preference is for males with the longest eye-stalks, and the bigger the male, the more females he has – up to as many as a dozen. The large males patrol up and down the thread striking it with their abdomens and shaking from side to side. This imposing display often causes smaller males to flee, though the very smallest sometimes hide among the females and stay in the harem, seemingly masquerading as members of the opposite sex.

Meanwhile, the big males square up. A male stalk-eyed fly's body size is closely correlated to its eye-span, and so rival males size each other up relatively fast. Only males of nearly equal eye-span (and therefore body size) fight it out. But it's a fight that's highly ritualized.

The males face each other aligned with their eyes exactly in parallel and forelegs extended. Then they raise themselves with their wings spread, drop down into a crouch, beat with their abdomen by bending their legs up and down, and return to their starting position with forelegs spread in threat. The winner, usually the larger male, repeats his imposing display while chasing the loser from the thread. If males are very evenly matched in size, the display can go on for 20 minutes, often escalating into physical contact, with the two wrestling with their forelegs. Once in a clinch, it is hard for the males to extricate themselves, and such wrestling matches may result in injury to either legs or eye-stalks. Ritualized displaying is therefore the safest way to decide disputes.

Having banished smaller males from the roosting thread, the dominant male can rest in the knowledge that, in the morning, there will be few males to compete with. At the first hint of dawn, the stalk-eyed flies start to stir, and a period of frantic mating begins. The male tries to mate with all the females in his harem before they disperse an hour or so after first light. He will watch a female closely and then leap onto her. Large males copulate frequently. The smaller males that hid among the females during the evening attempt to mate, but they fare less well: females prefer to mate with males bearing the longest eye-stalks, and large males frequently interrupt matings by small males, chasing them off.

The evolution of such exaggerated eye-stalks now becomes understandable. With their amazing eyesight – as acute as a dragonfly's – stalk-eyed flies can, from a distance of up to a metre (3 feet) away, accurately estimate the size of an approaching fly by his eye-stalk length and so assess a potential rival or suitor. The length and span of the stalks function as an instant signal to both males and females of the strength and virility of their bearers, allowing quick decisions to be made and avoiding costly battles and bad choices.

Above *A male stalk-eyed fly (second down) and his harem of females on their roosting rootlet.*

Opposite *Two males with equal eye-stalk width and therefore body size engaging in ritual battle to decide who has mating rights over the females. One rises up with forelegs stretched out in threat, while the other crouches down and drums his abdomen on the rootlet.*

Crippled by sex

The theory of sexual selection was first proposed by Charles Darwin in his book *The Descent of Man*, published in 1871. He also discovered the insect world's finest demonstration of this theory: the Chilean stag beetle, also known as Darwin's beetle. He came up with his theory to explain the extreme behaviours and physical shapes that just one sex of some species display. Male peacock tails and the odd shapes, colours

attracting a mate – only just outweighs the disadvantage of having an otherwise useless encumbrance. Beetles provide some of the best examples. The giraffe-necked weevil of Madagascar is one of the oddest. The male has an enormously elongated neck with a tiny head perched on the end, making the beetle unstable. It uses its neck in a kind of low-key combat, and females mate with the winners. But it was in 1835, when HMS *Beagle* was

and dances of birds of paradise are classic examples. They come about because the females of a species have, over countless generations, chosen to mate with males exhibiting most dramatically a particular behaviour or feature. This selective breeding has produced the peculiar animals we see today. Insects, with their flexible, mouldable exoskeletons are perfect for extreme sexual selection. In some cases, a selected feature becomes so extreme that the advantage of possessing it –

surveying the coast off Chile, that Darwin found his stag beetle, the male of which has reached such a peak of exaggeration as to be almost absurd.

There are more than 1000 kinds of stag beetle around the world, and the males of many have mandibles so enlarged they are useless for feeding. Instead, they are used for fighting. The Chilean stag possesses the most dramatic of all. They are at least as long as the rest of

its body and, whereas most stag beetles' jaws are elongated out in front, the Chilean stag's are curved down like scimitars. Its front pair of legs are elongated to raise its body up so the jaws don't snag as it walks, a solution that's only partially successful. One of the beetle's strongholds is around the beautiful Lake Todos Los Santos in the Patagonia region of Chile, surrounded by wooded slopes and overlooked by the snow-capped Orsono volcano. The male beetles fly unsteadily through the trees with a loud buzzing or wander on trunks and branches searching for the short-jawed females that inhabit the treetops. When a male ends up on the same branch as another male, in Darwin's words 'he is bold and pugnacious; when threatened he faces round, opens his jaws and stridulates loudly.' The beetles trundle towards each other, jaws open, and lock together.

Despite the size of their mandibles, fighting is not about injuring the opponent. As Darwin wrote, 'the jaws are not so strong as to cause pain to a finger.' The odd shape of the jaws is a design to allow a beetle to get them below the level of his opponent's wing-cases. The tip of each mandible forms a hook precisely the right size to fit under the wing-cases. The first male to

latch on to the other stands an excellent chance of winning the fight. His aim is to lever his opponent from his grip on the bark and throw him out of the tree. The male with the best grip pulls upwards, his long jaws giving great leverage. His opponent holds tight, his curved claws hooked into the bark. His legs pull out straight and the bark he is holding onto may even crack from the force. Battles can last from just a few seconds to several minutes. There are periods of calm when they hold each other in a clinch, like exhausted boxers, before one pulls free and begins to fight again. Once a male succeeds in levering his rival from the bark, he holds him high, leans over the branch and opens his jaws. The loser's mandibles slide down his aggressor's until he is holding on by the tips. Finally gravity wins, and he falls away down the tree.

A male may have many battles before he finds a female. But she frequently runs away. Even after he's caught up with her and persuaded her to mate, his fighting instinct remains, and he may pick her up and lob her out of the tree. Fortunately this is more of a short cut than a disaster. She needs to get to the ground to lay her eggs in the grass, where her grubs will feed on the roots.

chapter 5

Frogs, serpents and dragons

blood and banished to harsh and marginal environments by birds and mammals…
It would be easy to believe such ideas about reptiles and amphibians, as if they
were somehow inferior to today's dominant animal groups. It's true that the story of
their global rule began and ended long, long ago, but what a glorious history.

Modern amphibians are descended from the first vertebrates to leave the water.
Their fish ancestors had bones in their fins that became amphibian legs, and they
had lungs, probably because they lived in oxygen-poor swamps. Skin is so rarely
preserved in fossils that it's not possible to know when modern amphibians first
developed permeable skin for breathing. When those pre-amphibians climbed onto
land, they began a rule that lasted millions of years, producing a bizarre variety of
amphibious monsters, some as large as crocodiles. When the reptiles took over, their
great age lasted three times as long as mammals have so far dominated the planet,
reaching its peak with the dinosaurs. It took the crash of a meteorite to end this great
episode in life's history, leaving behind little more than ruined skeletons set in stone.
Only then did rat-sized mammals emerge from the safety of a nocturnal lifestyle to
spread and diversify, accompanied by birds, the direct descendants of dinosaurs.

Yet today's reptiles and amphibians are not ancient losers, struggling to compete in
the modern world. They are thoroughly modern creatures. Biologically, they are so
different from mammals and birds that they must tackle the problems of living in
different ways, but in many respects they are just as successful. They show an
astonishing flexibility both in behaviour and in physical shape and form, which
allows them to compete effectively and even to dominate some environments.

Modern amphibians are grouped into frogs and toads; salamanders (including newts
and mud puppies) and earthworm-like caecilians. They might seem disadvantaged
by their moist, permeable skin, their need to return to water or damp places to
breed and their cold blood, which leaves them needing to gain warmth from their
environment. But while their greatest diversity is in the wet tropics and temperate
zones, a dozen species of frogs, toads and salamanders survive at high altitudes and
latitudes through freezing winters. Their trick is to release glucose or glycerol into
their blood, lowering the freezing point of the water in their cells. In other regions,
the amphibians' cold blood is a clear benefit. Warm-blooded animals must feed
regularly, but cold-blooded ones can slow their metabolism almost to a stop and
wait out difficult times in suspended animation. This technique is spectacularly
demonstrated in the frogs that have invaded desert regions. In between the very
rare rains, these frogs burrow underground and cover themselves in an impermeable
layer of mucus. They can remain in this state for years. When rain finally arrives,
the frogs dig themselves out, sometimes in prodigious numbers, for a sudden and
rapid burst of breeding.

Above *A striped tree frog in a
Costa Rican rainforest – a
perfect environment for an
amphibian, which needs to keep
its skin moist at all times.
It displays camouflage colour on
top but bright yellow underneath,
which it reveals to signal that it
may not taste good.*

Opposite *An agamid lizard from
Borneo. A constant tropical
temperature allows this and
many other tropical reptiles to
be active throughout the day all
year round, unlike those in more
northern and southern latitudes.*

Previous page *An American
alligator – an ancient but highly
successful reptile from a dynasty
that was around long before
mammals.*

Reptiles are even more flexible and successful. Their bodies have evolved into an array of shapes and forms. They range from the 2.7-metre (9-foot) Komodo dragon and the saltwater crocodile, reputed to reach lengths of 6 metres (20 feet), down to a miniature gecko less than 2cm (0.8 inches) long living in leaf-litter in the Dominican Republic. They are grouped into the crocodilians; lizards, snakes and worm-lizards (such as amphisbaenians); turtles, tortoises and terrapins; and New Zealand's ancient tuataras. Some have developed points and spines to ward off predators. Snakes have lost their legs, multiplied their ribs and developed venom. Perhaps the most peculiar of all are the chameleons, with their wonderful ability to communicate moods and intentions through colour, their explosive tongues and unique split feet, which grip branches like callipers. Skin is a key to the success of modern reptiles. It is waterproof, and in combination with cold blood, it allows them to flourish in some of the world's most arid environments, where mammals and birds are just temporary visitors.

Reptiles and amphibians are not defenceless against attack. Frogs can puff themselves up and scream. The skin of poison-dart frogs contains some of the most toxic substances known. Some lizards run on water to escape. Snakes can inject or spit venom. The group's breeding strategies are hugely varied to cope with every kind of challenge. Frogs have sophisticated vocal communication to attract mates. They can also jam each other's calls, mimic other species and throw their voices like a ventriloquist. Some frogs and lizards care for their young as well as any mammal.

Yet perhaps the simplest way to judge the success of reptiles and amphibians is by their sheer diversity. There are 4500 kinds of mammals, 10,000 birds and 13,000 species of amphibians and reptiles. This group is not a remnant of a glorious past but an enduring success story.

Above *A Vietnamese pit viper in camouflage position on a branch, awaiting prey. Its strategy is to ambush, grab and inject venom, deadly to frogs, reptiles and small mammals. The development of venom is one of the reasons snakes have become so successful, allowing prey to be subdued without a great struggle.*

Left *A Vietnamese mossy frog, with perfect camouflage for its moist, shady environment. Amphibians have no external armour but have made use of the mouldable and colourful nature of their skin to develop disguises or warning signals.*

There still be dragons

Right *A Komodo dragon displaying its muscular, heavily built physique, designed for the short, powerful dashes needed to ambush large prey. It has come from a mud wallow, where it has attacked a feral buffalo, biting its back leg. The dragon's venom will prevent blood clotting, leading to bleeding. More bites over weeks will eventually kill the buffalo.*

It's been 65 million years since reptiles ruled the planet. Mammals and birds now dominate most natural habitats, but we can still get a glimpse of what the distant past must have been like. In 1912, a party of pearl fishermen working the treacherous waters of a remote Indonesian archipelago spotted giant carnivorous lizards patrolling the beaches of a small island. Their reports were the first that western science had heard of the Komodo dragon. Its home is just five arid islands east of Bali, at the point where the Indian and Pacific oceans meet – the smallest range of any large carnivore on Earth. Just a few thousand dragons now survive, with the biggest population on Komodo Island itself.

The dragon has the charisma of a top predator, a combination of power, intent and armoury. It rules its tiny world just as surely as its ancient dinosaur relatives ruled the planet. The Komodo dragon owes its continued existence here to the fact that no top mammal predator was ever able to survive in such an austere place. Prey is rare, and a warm-blooded predator needing to feed every few days would rapidly wipe it all out, leading to its own extinction. But the cold-blooded Komodo dragon only needs to feed a dozen times a year. As a result it has thrived.

The dragon achieved lasting fame not just for its remarkable size (average size of a male, including its tail: 2.2m/7.2 feet; average weight: 80kg/176 pounds) but also for its fearsome reputation as a predator with the ability to hunt humans. Usually it employs a classic sit-and-wait strategy, spending days motionless beside a forest track, waiting for prey such as deer to pass. Then it explodes out, reaching speeds of 18kph (11mph) and landing a huge bite on the prey's throat or belly, causing terrible wounds with its 60 sharp, serrated teeth. The prey is quickly overpowered and devoured on the spot.

This fate has also befallen human visitors, most famously the Swiss Baron Rudolph von Redling in

1974. The 84-year-old adventurer chose to separate from his party on Komodo Island and follow a different path back to the shore, where a fishing boat was waiting. He never turned up. Despite the efforts of a 100-strong search party, the only remains ever found were some pieces of his camera.

The dragon is an efficient predator of smaller prey, but what makes it such a success on these islands is its ability to tackle animals much larger than itself.

Its original prey, believed to include a species of miniature elephant, has long since disappeared, but their place has been taken by introduced animals, including the water buffalo – a dangerous grazer many times the size and weight of a Komodo dragon.

The dragon's method of hunting buffaloes is one of the most gruesome in the reptile world. Prime time for hunting is the season when the islands are burning hot and dry. The last water is concentrated in a few waterholes, which the buffaloes must visit daily. The dragons wait for them there. A hungry or aggressive individual will try to sneak up behind a drinking buffalo and bite a leg or its genitals. Then it rapidly backs off to avoid the furious mammal's horns.

The tissue damage from the bite is not enough to kill. Until recently, it was thought bacteria in the dragon's saliva poisoned its prey. But it has now been shown that the dragon, like some snakes, has venom, making it the world's largest venomous animal. The venom rapidly kills small prey by preventing blood coagulation, leading to massive blood loss. It also lowers the prey's blood pressure, sending it into shock. Repeated bites on large prey such as a buffalo will eventually kill it. During this time, the buffalo's wounds become infected. A Komodo dragon can detect decaying flesh from up to 6km (3.7 miles) away and homes in on

the injured animal. Seven or more individuals might track it each day as it walks to and from the waterhole. They try to land more bites as it weakens. Eventually the stricken creature collapses, and the dragons overwhelm it, beginning to devour it even before it dies.

Dragons feed on a strictly largest-first basis. Big dragons chase off smaller intruders, sometimes killing and eating them. A hungry dragon rips off great chunks of meat, swallowing bones, hooves and all. It can eat up to 80 per cent of its body weight in one sitting. A group will strip a carcass to the bone in just a few hours, leaving only scraps for smaller dragons. The fully fed dragons must then spend hours lying up in the sun, keeping their body temperatures high to speed digestion, otherwise the food will rot inside them.

The dragon's mating rituals are also striking and brutal. Males wrestle for access to females. They sling their front limbs around each other and rise up, eventually balancing on just their tails as each tries to overpower the other. Their 10cm (4-inch) claws rip through skin, leaving bloody wounds. Each fight is short, just a matter of seconds before one or both crash to the ground in a cloud of dust. But confrontations between two evenly matched dragons may continue on and off for days, until an exhausted loser finally backs off.

Beating off rivals, though, is only half the battle, because females are rarely receptive. A winning male tongue-flicks a female to gauge whether she is in the right stage of fertility. He may also scratch her back or rub his chin on her skin to stimulate her. When he tries to mate, she frequently resists him with her teeth and claws. The male then has to use his full body weight and powerful limbs to restrain her. Curiously, for an animal with such a violent courtship, the Komodo dragon is one of the few lizards that are monogamous – having only one partner – and forming close pair bonds.

The mated female digs a burrow in an abandoned, sandy mound originally built by the orange-footed scrubfowl to incubate its own eggs. She lays up to 20 eggs and stays near the mound for the 7 months it takes the eggs to hatch. But despite such effort, many of the young dragons – a foot long (30cm) when they hatch – are cannibalized by their own kind and can make up 10 per cent of the diet of large dragons. For their own protection, they therefore spend much of their time up trees, where they hunt for insects and smaller lizards.

The Komodo dragon is far from being a dinosaur. It is a modern reptile, but its domination of its habitat mirrors that of its ancient ancestors in a way that, today, is unique.

Next page A dragon feeding on a buffalo that succumbed to its venomous bite. It produces copious saliva, probably to make swallowing the dry parts easier. The prey was tracked down by the smell of its putrefying wound and attacked and devoured by ten Komodo dragons.

The lizard that runs on air

Opposite An Asian Sail-tailed Lizard using the same water-sprinting technique that basilisk lizards use. Like them, it can also escape by swimming and holding its breath for long periods under water.

Below Foot-work in action. The basilisk's rapidly windmilling feet create air pockets as they slam down, and that is what they are running on. It's very energetically demanding, though, and if the water expanse is large, the escaping lizard runs out of energy and resorts to diving.

The basilisk of Greek legend was a fearsome creature. King of the reptiles, it wore a crown and possessed supernatural powers, including being able to kill at a glance. The basilisk lizard of Central and South America has a crown-like crest. And the resemblance to legend doesn't stop there. It also has a remarkable talent (only recently explained) for escaping predatory snakes, birds and mammals. In moments of danger, it can sprint on water — hence its nickname, the Jesus Christ lizard.

The basilisk (actually five related species) is found in tropical forests from Panama to Ecuador. It spends a lot of the day sitting stone-still on branches from where it can spot predators, and shows a preference for living near water, the reason for which becomes clear when something startles it. The basilisk flings itself into the air. If it doesn't fall into water, then it runs to it as soon as it touches ground. It skitters over the water surface on its hindlegs at 1.5m (nearly 5 feet) a second, forelimbs windmilling, and keeps running when it reaches land. But on large expanses of water, it runs out of energy and dives, staying submerged for up to half an hour.

For tiny, light creatures such as insects, water tension is like the membrane of a balloon, which they can walk on without puncturing. But the basilisk lizard, technically, should sink. All the more remarkable is the true reason for its ability. The Jesus Christ lizard was misnamed. Researchers have discovered that it

doesn't run on water at all. It runs on air. It slaps its hindlegs in turn on the water surface and drives them down, creating air-filled potholes in the surface, its feet dry within. The feet push back and out, giving forward motion and stabilizing sideways wobble. Then the masterpiece of creating an air hole is revealed. The lizard's feet move so fast they can complete a cycle and lift back above the water surface before the air pocket collapses. If the legs pushed through the water, the drag created would be so great that the lizard would sink. Yet its tail trails behind through the water, seemingly undermining the whole system. In fact, it probably acts as a counterbalance, preventing the lizard from falling flat on its face.

This water-running appears graceful. But seen in slow motion, it is ungainly and obviously exhausting. Frequently the lizard trips over the water surface and lands with a belly flop. Then it has to dive. A newly hatched basilisk lizard seems to bounce across the water surface. But water-running becomes ever harder as it grows heavier, its feet become proportionately smaller and its running speed proportionately slower. An adult lizard weighs about 200g (7oz), which is near the limit at which it can generate enough power to prevent itself sinking. By comparison, for a human to run on water would mean driving the legs down and back at 105kph (65mph) — a muscular exertion 15 times greater than a mere human is capable of.

Lost worlds and bouncing toads

Above *The Kukenan pebble toad on top of its mountain home, revealing its Mickey Mouse hands with their flexible toes. Every flat-topped mountain, or tepui, that has been explored has its own species of pebble toad.*

Opposite *The top of Kukenan, where it rains almost every day and the vegetation is dominated by insect-eating plants. Surprisingly, the Kukenan pebble toad can't swim, but then it doesn't need to, as the air is humid enough to keep its skin moist out of water.*

At the peak of the great age of reptiles, 180 million years ago, dinosaurs roamed over a vast sandstone plateau that overlaps modern-day Venezuela, Brazil and Guyana. Over the millennia, the plateau has been fractured by movements of the Earth's crust and eroded, millimetre by millimetre, by the daily action of water on rock. The result is a startling landscape unlike anywhere else.

This region – the size of Texas – is littered with more than a hundred mountains called tepuis (meaning House of the Gods), each a remnant of that ancient plateau. They jut, sheer-sided and flat-topped, out of a sea of rainforest. Some are nearly 2km (more than a mile) high and generate their own weather systems. Clouds well up the sides, like water running against gravity, before pouring over the top. Sir Walter Raleigh, on his

expedition up the great Orinoco River in 1596, was the first to send reports of this place back to Europe. Many were sceptical, believing such a landscape could only exist as a fantasy. But in the 400 years since then, this region has captured the imagination of generations of explorers, scientists and romantics.

It's a source of legends of riches. Raleigh believed it was the gateway to El Dorado – the mythical city of gold. Jimmie Angel, discoverer of the Angel Falls – the world's highest waterfall, plummeting off Auyan tepui – even claimed to have found nuggets the size of a fist in a river. But it was the British novelist Arthur Conan Doyle who came closest to identifying the real wealth of this region. In 1885, his imagination was fired by a report of the first ascent of a vast tepui called Roraima.

Right *The amazing, bouncing pebble toad escaping from a tarantula. It tucks its legs in and just bounces down the rocks, coming to a stop unscathed.*

Opposite *The pebble toad's close relative, the waterfall toad, living in rainforest below the mountains. It reveals what those odd feet were originally designed for. When attacked by a snake it simply lets go and falls, but on the way down, it grasps a passing leaf or stem with its long, strong toes, hangs on and then pulls itself up to safety.*

The result was his classic adventure story *The Lost World*. He described an isolated ecosystem dominated by dinosaurs, surviving on the flat top of a mountain. Conan Doyle was wrong in the details but correct that the riches are biological. Each tepui is like an island, on which species have followed independent evolutionary pathways. This is why some call it the inland Galapagos, after the island group that so influenced Charles Darwin. But the tepuis are far less known than Galapagos and remain remote, even today. To reach one on foot requires a long trek through virgin forest cut by treacherous rivers. Some locals refuse to act as porters because of the *mepepire*, or fer-de-lance – the most feared snake of the region on account of its readiness to bite and its toxic venom. This leg of the journey is, however, the easy part. The sheer sides of the tepuis, the biting insects and the constant rain make them uniquely difficult to climb. Even so, it is remarkable that most tepuis have still not felt the tread of explorers' feet.

The few biologists who have ventured onto tepuis may not have discovered Conan Doyle's dinosaurs, but they have encountered the pebble toads – a different species on each tepui explored. They demonstrate how even vulnerable amphibians can outwit predators. Measuring less than 3cm (1.2 inches) from end to end, these tiny toads are black, knobbly and peculiar in several ways. They can't swim and they can hardly hop. Their feet seem oversized and have flexible digits, like Mickey Mouse hands.

The best known is the species living on a tepui called Kukenan ('House of the Dead' to the locals, who believe spirits of the dead reside there). It spends its time walking very slowly around its otherworldly landscape. The sandstone is riddled with sharp quartz crystals, fissured and potholed, and for more than $2km^2$ (1.2 square miles), it's eroded into thousands of arches and haunting shapes. There is no soil, just rock and little 'gardens' containing unique carnivorous plants. Visitors describe the silence behind the wind, broken only by an occasional bird or the piping of the toads.

A pebble toad appears defenceless, and yet it survives in the presence of predators: a scorpion, a tarantula and visiting birds. The way it avoids danger is perhaps the most extraordinary thing about it. It simply lets go and rolls away, bouncing from rock to rock, rigid as a rubber toy. It may come to rest when it hits flat ground, or it may somehow stop on an almost vertical surface. Danger gone, it continues its slow perambulations.

This bizarre behaviour raises interesting questions. Where did these toads come from, and how did they develop this peculiar behaviour? American scientist Bruce Means has spent 20 years exploring the region, often doing it the hard way, on foot. He is known for an uncanny ability to find animals and recognize new species, and the tepuis have provided rich pickings. His rediscovery of a toad last seen 100 years ago and known only from a few specimens in the Natural History Museum, London, may have provided some answers. He calls it the waterfall toad after the beautiful location where he found it on the lower slopes of Mount Roraima, Conan Doyle's inspiration.

This toad has the same odd feet, but Means discovered it uses them in a new way. It prefers to sit on leaves, and if it senses danger, such as the approach of a snake, it lets go and just falls. But now its feet come into play. As it tumbles, it holds out its arms and legs and grabs any passing twig or leaf. Its ability to grip is astonishing, and more often than not, it will end up dangling by one foot and then slowly haul itself up. Means suggests that the hands and feet of this group may have evolved in the waterfall toad or its ancestor to enable it to move along and hang onto leaves and twigs, rather than hop to safety. Then, as each tepui became isolated, toad populations became adapted to living high up on them, using their grasping hands and feet to grip vertical rock faces and stop themselves in mid-roll.

To the experts, the tepuis ecosystem is like a diamond, most of whose facets have yet to be seen. These toads are but one glint – one tale of many yet to be told.

Mother and the dirty digger

Reptiles have developed all kinds of strategies for safeguarding their young from threats posed by creatures such as mammals and birds. Madagascar's collared iguana, for example, has what appears to be a fail-safe method of protecting its eggs. But while it might fool the warm-blooded creatures, a cold-blooded snake is another matter.

The collared iguana is the most abundant lizard in the western forests of Madagascar. It spends most of its time safely up a tree, resting on branches or even vertically on the trunk. It darts to the ground to grab insects and other invertebrates and then rushes back up again. When threatened by a potential predator such as a bird of prey, it retreats into a crevice and blocks the entrance with its spiny tail.

But when the first rains of the wet season arrive, the female has to go to the ground for a much longer time, to deal with the most important event of her life – laying her eggs. She slowly searches for exactly the right spot, usually a patch of bare, sandy soil. She digs a hole and lays a small batch of eggs, and then, like a dog burying a bone, she pushes the eggs with her nose right to the bottom before neatly covering it over. She leaves little indication that anything is there.

This strategy is usually enough to protect the eggs from mammals and birds, but the iguana has another problem. This comes in the form of another reptile, the hog-nosed snake, easily identified by its upturned nose.

When the snake sweeps its head from side to side, its nose makes a very effective digger. It is common in areas where the iguanas breed and can often be spotted lurking in the undergrowth or even lying brazenly out in the open while the iguanas are burying their eggs.

Once the iguana has buried her eggs, or even while she's burying them, a hog-nosed snake may appear and immediately dig them up again. (It may have watched the iguana laying, but it also has an excellent sense of smell.) It rapidly swallows them whole, sometimes several at a time, and the iguana is helpless to do anything other than just watch the round bulges move down inside the snake's body.

It would be easy to view the iguana's helplessness as a kind of failure. But offspring survival is a numbers game. The iguana has developed a good enough strategy when it comes to most predators, and not every mother iguana will have her eggs raided.

It's also possible to see the situation from the snake's point of view – a reptile that has been flexible enough to find a way around the iguana's sensible strategy of hiding its eggs underground.

Know your predator

Top *Taking defensive action against a coachwhip snake. The female regal horned lizard can't outrun the snake, but it has another strategy. As the snake approaches, she turns side-on, raises her horns and tilts her back to look like a huge mouthful, then quickly flips over to reveal a wide, white, apparently spine-lined underside. It's enough of a shock to make the snake flee.*

Opposite *Preparing the nest hole. Once the eggs have been laid, the female will guard them for a week or two. Her range of defences are tailored to the predator. Should a coyote or dog threaten, she will squirt a jet of distasteful reptilian blood from the corner of her eyes.*

The regal horned lizard lives in desert shrub habitats of the southwestern United States. It faces the same problem as the Madagascan collared iguana: its eggs are dug up by an egg-eating snake. But unlike the collared iguana, the horned lizard does something about it. It displays the kind of behavioural flexibility that underlies the success of modern reptiles. Scientists Wade Sherbrooke and Clayton May have recently discovered that it is unique among reptiles, first in its ability to distinguish a range of predators and second in having a different defensive strategy for each of them.

Instead of laying a clutch of eggs and then burying and abandoning it as a collared iguana does, the female regal horned lizard stays near her clutch for up to two weeks. By doing so, she puts herself at risk of being eaten and leaving her eggs as vulnerable as the iguana's. Her response to each kind of predator is tailored to give her and her eggs the best chance of survival.

While the greatest threat comes from snakes, the lizard is also eaten by wild members of the dog family such as coyotes and kit foxes. When threatened by such a predator, the lizard's response is to fill her sinuses with blood and squirt a jet from the corners of her eyes into the predator's mouth. It seems the blood contains distasteful substances, and the dog beats a rapid retreat. The lizard can deal with three kinds of snake. If it spots a rattlesnake – which is too slow to chase prey,

relying instead on sit-and-wait and its venom – she simply runs away. The coachwhip snake, though, is a different proposition. It isn't venomous but is just as dangerous as a rattlesnake because it chases prey at surprising speed, and a horned lizard can't outrun it.

A coachwhip must judge whether its prey is the right size, since it can't swallow outsized prey the way that a python can. Attempting to eat something too big or the wrong shape could be fatal. So when the horned lizard sees a coachwhip approach, she turns side-on and tilts her back towards it by pushing up her legs on one side only, which makes her appear taller and larger. She also raises the horns behind her head to emphasize what an uncomfortable mouthful she would be. Finally, she flips onto her back and lies still. The camouflaged topside is now replaced by the almost pure white underside. Her limbs stick out stiffly, and the fringing scales running along her side give the appearance of spines. The whole process both startles the snake and enhances the awkwardness of the lizard's shape – often enough to prevent an attack.

The third snake the lizard reacts to is the one that wants to eat her eggs rather than her. But western patch-nosed snakes are relatively small, and so the lizard goes on the attack. She rushes towards the snake, butting and biting it. The snake, shocked by the ferocity of her assault, turns and races away.

The Reactolite chameleon

Chameleons are odd. Even reptile experts say so. Their cartoonish appearance and unique lifestyles are a step apart from the rest of their group. They are a beautiful example of the physical flexibility that has allowed reptiles to invade so many environments. The Namaqua chameleon is the exception that proves the rule. It is peculiar even among chameleons because it has overcome the hindrance of a body precisely adapted to one kind of habitat and has invaded an entirely different one.

Above *A Namaqua chameleon with a dune beetle meal. Its prey runs fast across the hot sand, and instead of lying in wait for insects, which would mean starving, it runs after them. But like a normal chameleon, it fires its projectile tongue to snap them up.*

Right *Warming up. In the early morning, it turns dark on one side to absorb the sun's heat and white on the other side, to help prevent heat escaping – an extraordinary and very chameleon-like adaptation to desert life.*

Almost all chameleons are found in Africa and Madagascar, though a few reach into Asia and southern Europe. They come in two forms. There is a group of tiny ones – the Brookesia chameleons – found only on Madagascar, which live in the leaf-litter. But the other 130 species, all much larger, are perhaps the more remarkable, since they are so superbly adapted to living in a very specific kind of habitat. While other animals inhabit places broadly described as mountains or caves or the deep sea, the best description of these chameleons' home is… twigs.

Their bodies and behaviour are radically different from other reptiles. They move slowly, often rocking backward and forward to mimic the movement of a leaf. Their feet are split like pincers, making them brilliant at gripping twigs and branches, but they're much less useful on their rare journeys to flat ground. They are far too slow to chase prey. Instead they wait for something to land nearby and fire their extending tongue at it.

These reptiles are so specialized to their environment that they appear to have hit a biological dead-end. It seems impossible that they could ever adapt to living anywhere else. Yet the Namaqua chameleon has done exactly that – and thrives in an environment that could hardly be more different: a place with a name translating as 'open space'. It is the Namib Desert – a vast band of gravel plains and huge sand dunes running down the Atlantic coast of Namibia.

The challenges here could scarcely be greater. The chameleon has twig-gripping feet, but the ground is flat. Chameleons are slow, but the most abundant food here is a beetle that runs at tremendous speed to save burning its feet on the baking sand. Chameleons usually live a skulking, shaded life, but there is no shelter here from the heat or from predators. Chameleons are intensely solitary but usually quite abundant – so it is inevitable that they will meet and breed. Yet the Namib Desert is vast, and the Namaqua chameleon is scarce.

The chameleon has probably been here for a very long time. The Namib has existed for 55 million years, making it the oldest desert on the planet. This has given enough time for all kinds of animals to evolve special ways of dealing with such a harsh place. There is a spider that kills ants by pulling them onto the burning surface of the sand. A lizard that balances on two legs when standing still, keeping the other two off the hot sand. Snakes and moles swim through the sand. A beetle allows sea fog to condense on its back so droplets can run into its mouth.

The Namaqua chameleon, too, has overcome the odds. Its feet remain split in the chameleon way, but when spread out on the sand, they form a wide, flat base for walking. It can sit still and wait for a passing beetle just as any other chameleon would, but unlike any other chameleon, it can also sprint after its prey. When facing its greatest threat, predatory birds scanning for food, the chameleon hunches up, presses itself onto the sand and turns a darker shade, looking for all the world like a small pebble. One if its most startling adaptations is the ability to be different colours on each half of its body, turning darker on one side to absorb the sun's heat early in the day, while keeping the other half of its body white to prevent heat escaping.

Perhaps its greatest challenge is reproduction. In such a vast place, a wandering male spotting a female has to make the most of a rare opportunity to mate. There is no scope for subtlety. He walks sideways towards her, flattening his body to make himself appear taller and more impressive while changing colours to a very contrasting pattern to show his state of excitement. If she doesn't show any interest, he runs at her and beats her into submission with head-butts and bites before he climbs on top to mate. Afterwards he might run after her and try to mate again and again until she finally gets away from him. It is a harsh technique but an essential tool in the chameleon's adaptation to one of the toughest environments on Earth.

Above *The mating chase. In such a harsh environment, chameleons are widely dispersed, and it might be months before a male finds a female. So when he does, he will mate with her whether or not she wants him to. This may involve violence and certainly a chase.*

Opposite *Sand-walking. The typical calliper-like chameleon feet, originally designed for holding on to twigs, are spread out flat to provide double the surface area so the Namaqua chameleon doesn't sink into the sand. It's just one way that this desert-living reptile has turned what should be a disadvantage into an advantage.*

The submersible serpent

Adopting an entirely marine existence presents exceptional problems for any land creature – problems of movement, predators, breeding, feeding, breathing. Sea snakes have come up with some beautiful solutions. One particular species in just one tiny speck on the globe tackles an apparently insoluble problem in a way that is both breathtakingly inventive and simple.

The move into the sea freed the snakes from competition with other land animals. Their sinusoidal slither pre-adapted them to life in water, and they became longer and slimmer so they could slice more easily through the water, their tails flattened like paddles to provide thrust. Yet these snakes still have neither the speed nor the agility to chase fish. This is where their most famous attribute comes in. Many sea snakes have developed venom of alarming toxicity, jostling with creatures such as box jellyfish, funnel-web spiders and stonefish for the title of the world's most venomous creature, though they rarely bite people and even more rarely kill them. Their venom is enhanced to compensate for their slowness, paralysing prey in seconds.

Breathing was a major challenge. A sea snake doesn't have gills and must break the surface to take a breath. But a lung the length of its body allows it to submerge for long periods, its windpipes closed and its nostrils blocked, its connection to the world of air reduced to an occasional breath. But for a few sea snakes, the

dependence on air and land remains. The problem is breeding. Sea snakes evolved from the elapids, a group that includes cobras and mambas, which lay their eggs under rocks or logs or in crevices, the eggs absorbing oxygen through their shells. But sea water doesn't hold as much oxygen as air, and so laying eggs under water isn't an option. Most of the 62 species of sea snake solve the problem by 'incubating' their eggs internally and giving birth to live young, but 5 species – the sea kraits – never made this change. Females have to lay their eggs on land, exposing themselves, the eggs and their newly hatched young to land-dwelling predators. Yet one sea krait, living around the tiny Pacific island of Niue, has solved the problem with wonderful ingenuity. It lays its eggs on land but in a place where no predator will find them – under the island.

Niue is the limestone tip of an undersea mountain, its soft rock riddled with caves. The snakes swim through an underwater tunnel to a cavern at the end of which is an air bubble and dry land. The snakes crawl up into crevices and ledges on the walls and roof, where they lay their eggs, and then leave. The eggs develop over several months, safe from land predators and kept moist by the fine mist produced each time the swell rises, compressing the air. When they hatch, the baby snakes slide and tumble into water, swimming out to the sunlit sea. This is as close as an animal dependent on land can ever get to being fully aquatic.

Above Mating Niue sea kraits. The male has grabbed the female while she was taking a breath at the surface and has twisted himself tightly round her to stop her escaping while he mates. She will try to dislodge him, even by swimming into tight cracks, and will swim away fast once mating is over.

Below, left A baby krait emerging from its leathery egg, laid in an air pocket above sea level in a cliff-side cave. The young would suffocate if the eggs were laid in water.

Below, right A baby making its way down to the underwater entrance of the cave.

Opposite Resting Nieu sea krates in Snake Gully, a favoured gathering place. Every 15 minutes or so the snakes need to surface to take a breath of air.

Next page A ball of resting snakes. To sleep properly, they must find a safe spot out of the water so they can breathe, often choosing a cave with an underwater entrance.

chapter **6**

Brilliant birds

Above *A snowy owl displaying the aerodynamic prowess that has made birds such a successful group. Each species' plumage is adapted for its lifestyle. In the case of owls, it provides warmth, especially for night flight, and wingbeats that are comparatively silent for hunting.*

Opposite *A chinstrap revealing wings adapted as flippers and plumage designed for a life in a cold sea. Its feathers are tightly packed and overlapping, providing a waterproof layer over downy undershafts that trap air for insulation.*

Previous page *A male goldeneye displaying. In addition to warmth and waterproofing, his colourful plumage provides finery for advertising health and prowess to potential mates.*

IT HAD THE HALLMARKS OF A REPTILE but bore the unmistakable imprint of feathers. It was the size of a pigeon, with a long, bony tail, jaws studded with teeth and forelegs each with three separate digits ending in a curved claw. What the quarrymen had discovered in the fine-grained Jurassic limestone in Bavaria was the earliest known bird, fossilized since the age of dinosaurs and now with the name *Archaeopteryx lithographica*, meaning 'ancient wings in slate'.

For more than 150 million years, birds have brought colour, beauty, spectacle and song to the planet. Today we can list around 10,000 species. The bee hummingbird is the smallest, weighing just 1.8g (0.6 ounces) yet able to beat its wings up to 200 times a second. By contrast, the wandering albatross, with its 3.5-metre (11.5-foot) wingspan, can soar for hours across open ocean without a single beat. Arctic terns fly up to 35,000km (21,750 miles) a year – more than a million kilometres in a lifetime. And emperor penguins 'fly' the bitterly cold seas of the Antarctic, diving to depths of 500m (1640 feet) and holding a single breath for up to 20 minutes. The ostrich has done away with flying, investing in size instead. And when it comes to avian extravagance, the colour and beauty of the birds of paradise takes some beating.

To become masters of the air, birds have overcome the obstacles that restrict the movements of land animals. They outfly the insects, which, though they took to the air some 200 million years before birds, have bodies constrained by exoskeletons and so cannot operate above a certain size. Wings allow birds to hover, soar, dive at speed, and fly upside down and backwards, and even to circumnavigate the world – reacting to changing seasons, feeding opportunities and one another.

What truly sets birds apart from other creatures is their feathers. These may have evolved from elongated, frayed scales on the trailing edge of ancestral forelimbs, useful for gliding or parachuting, or it may be that scales became featherlike to act as temperature regulation devices, particularly heat shields, enabling the ancestors of birds to be more active in hot habitats. Whatever the case, feathers were a design breakthrough. Made largely from keratin, the same protein in hair and nails, they are strong, light, warm and flexible. Their insulation helps maintain a body temperature of 40-42°C (104-107°F), they give power, lift and manoeuvrability, and their colours – created by both pigments and feather design – enable communication and camouflage.

Feathers and flight may set birds apart from other animals, but they face the same challenges of finding enough food, escaping predators, attracting mates and raising their young. What follow are stories of some of the more sophisticated ways modern birds have solved the eternal problems of life, whether it be lesser flamingos choosing to settle en masse on the caustic but safe location of a Kenyan lake, gangs of great white pelicans using brute force to plunder Cape gannet nesting colonies, or male Vogelkop bowerbirds using bower decoration rather than fine plumes to get a mate.

The caustic choice

Africa's Great Rift Valley stretches for more than 6000km (3700 miles), and along its length lies a necklace of lakes. Some of these have so little outflow that they have become giant cauldrons of caustic soda, saturated with alkaline mineral salts where ground temperatures can reach a blistering 60-65°C (140-150°F). Yet one bird has learned how to exploit this extreme environment in surprising and spectacular fashion – the lesser flamingo. Its name is derived from the Latin *flamma*, meaning fire, and it is sometimes known as the flamebird.

Lake Bogoria in Kenya is flanked on one side by a dramatic Rift escarpment and on the other by steaming vents gushing boiling-hot water up into the sky. Lesser flamingos visit this lake to harvest spirulina, a type of cyanobacteria (often referred to as blue-green algae) that periodically multiply explosively in the warm, carbonate- and phosphate-rich water, transforming it into a nutritious pea-green soup. Flamingos are the only birds that properly exploit this valuable resource, filtering the surface water with highly specialized curved bills, which contain up to 10,000 thin, sieving plates – 'lamellae'. The birds' feet stir the water, their heads swing from side to side, and their tongues move in and out like pistons – 20 times per second, filtering up to 20 litres (35 pints) of water a day to net a precious 60g (2 ounces) of spirulina. The carotenid pigments in the bacteria give the flamingos their colour, turning blooming lakes into a riot of pink.

In a productive year, more than a million lesser flamingos – about a third of the total population – gather on Lake Bogoria in Kenya to harvest this bumper bacteria crop. To quench their thirst, tens of thousands of adults queue and jostle their way to a few key freshwater river mouths to wash and drink. The younger, paler immature birds are forced out on the periphery, often right along the lakeshore, making them highly vulnerable to predatory assaults by olive baboons and African fish eagles. Marabou storks also deploy a cunning strategy, walking along slowly and

Above *Lesser flamingo displaying one of the most specialized of all beaks, designed for sieving spirulina (cyanobacteria) from warm, caustic water that most other birds would find toxic.*

Right *The spectacular flamingo gathering at Lake Bogoria, Kenya. The birds come here to feed on the blooms of nutritious spirulina. Pigments in the spirulina give the flamingos their pink colouring.*

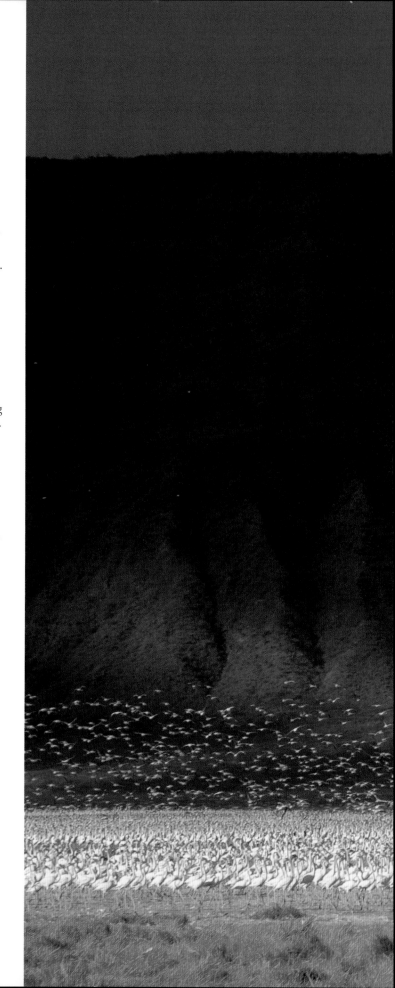

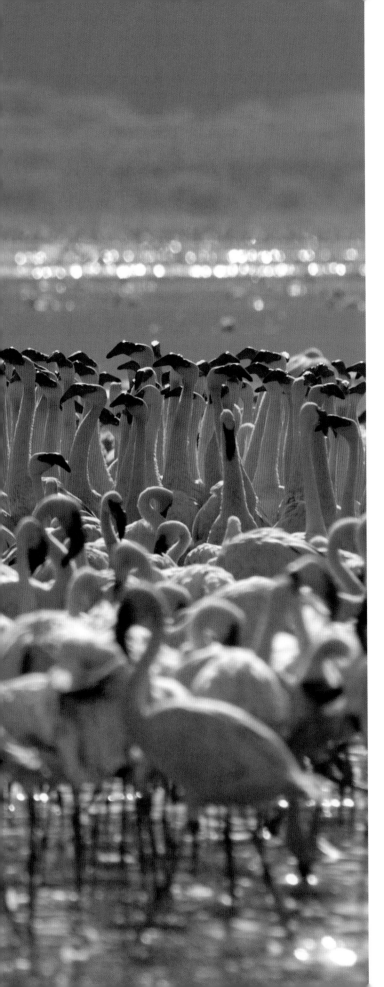

'flushing' the flock to see if there are any weak or injured stragglers, before hammering them with their powerful beaks.

When feeding conditions on the lake are good, they can trigger one of the most spectacular of wildlife events – a dazzling courtship dance. With wing flashes, head flicks, beak nibbles, neck moves and very distinctive vocalizations, groups of flamingos begin marching, and as the momentum grows, more and more join the parade. So smooth are the rapid footfalls and body moves that it appears as if the birds are gliding on water. The fluffed, outstretched necks of the dance party are also pinker than normal, giving the flamingos a very distinctive look that suggests they are approaching breeding condition.

In a seething, spectacular mass, the group constantly splits, reunites and changes direction, helping bring reproductive synchrony to those involved. What it is that actually causes a pair to bond – and perhaps to mate later – is not clear. Is it the head height, the calls, the feather colour and health, the number of neck and wing moves or perhaps a combination of one or more of these things? Whatever the case, this great parade is a magical prelude to nesting.

Specialization has its perks but also its problems. Spirulina is a highly nutritious resource, but it's also very unpredictable. Algal blooms can die off as swiftly as they arise, and so the lesser flamingos must remain opportunistic nomads, constantly shifting lakes and travelling in the dead of night, to exploit the best the Rift has to offer. Suitable feeding lakes can be found almost the length of Africa, from Ethiopia in the north to the salt pans of Namibia in the south, but there are only two favoured nesting sites – Lake Natron in Tanzania and Lake Magkadikadi in Botswana. This could be because the lakes offer remote and relatively safe nesting options, with reduced predator pressure and, in decent years, a fairly reliable food supply. But no one really knows. What we have learned is that

Above *Fish eagles fighting over a flamingo picked off at the edge of the colony. Eagles and marabou storks take the young, the sickly and the stragglers.*

Left *Starting the dance promenade. With heads held high and feathers ruffled to look as pink as possible, a group begins to strut, attracting more and more to a dance that leads to pair formation.*

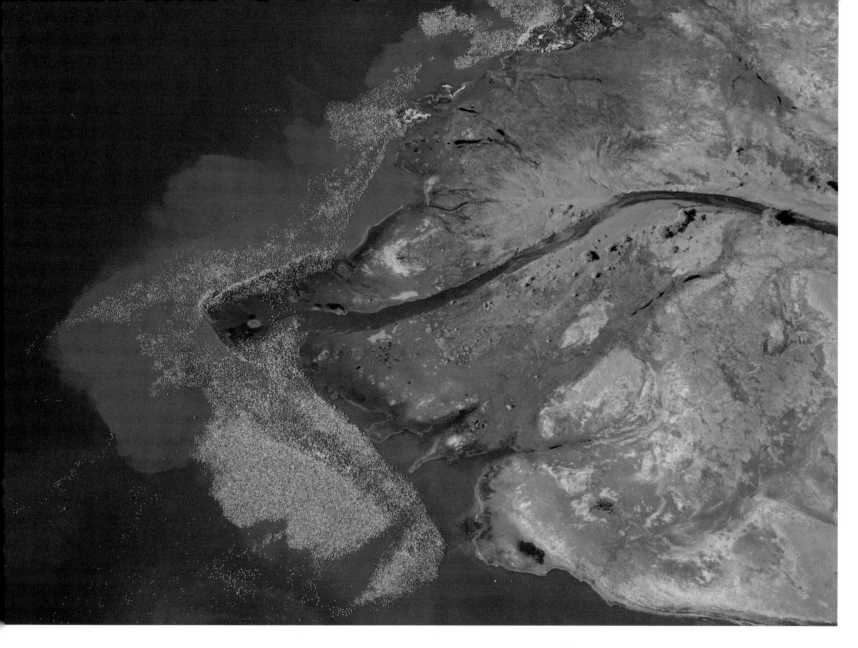

Above *A tiny proportion of the more than a million lesser flamingos that gather on Lake Bogoria to feed.*

Opposite *A huge congregation feeding in the shallows, relying on safety in numbers, the barrier of caustic water and the power of flight to avoid most predators.*

breeding is rarely an annual event on these lakes. Sometimes five or six years can pass without success. When nesting kicks off, the flamingos build cone-shaped mounds with globules of mud just a little way off the ground, where it's a bit cooler and there's less chance of flooding. A single egg is laid. If it hatches and the chick survives to join a crèche, there is no guarantee it will reach adulthood, because the lake can be as fickle as the supply of spirulina.

In the wettest years, nests can flood and chicks drown, and in droughts, chicks can perish from heat stress and starvation. Those that get through this perilous start may find that, to get to the nearest food and water, they must make a treacherous march over many kilometres of baking-hot, sticky soda. Given safety in numbers and food provided by their parents at night, they have a fighting chance – if they can avoid getting shackled with anklets of soda or assaulted by predatory lappet-faced vultures. It's a miracle that any make it at all.

Recently there have been new and perhaps even greater threats to flamingo success on these crucial lakes. Proposals have been made for soda-ash mining and hydroelectric power on Lake Natron, which could bring major risks through disturbance, increasing numbers of nest predators and changing water balance and lake chemistry. Elsewhere land claims, water pollution and human disturbance are all growing concerns, threatening the spectacular life of this extreme specialist.

The knot's clock and pitstop

Above *Horseshoe crabs coming ashore in Delaware Bay to spawn, awaited by hungry red knots eager for the feast of eggs. It's a crucial rendezvous for the avian migrants on their marathon trip, and precise timing is vital.*

Opposite *Red knots devouring the tiny eggs coating the pebbles. In a matter of weeks, the birds will have almost doubled their weight and be ready to fly on to their Arctic breeding grounds.*

The red knot is one of the champion long-distance migrants of the bird world, flying almost the entire length of the planet, not just once but twice, in a year. Food availability is what sets its clock, and timing is everything.

Between mid-March and mid-April, as the southern winter sets in, this little shorebird leaves its feeding grounds in Tierra del Fuego, Chile and Argentina, to fly north to its breeding grounds in the Canadian Arctic. It's a journey of almost 17,000km (10,000 miles) – each way. For a bird with a wingspan of just 50cm (20 inches), this is a massive feat, and success depends on careful timing and crucial seasonal staging sites.

The first stopover for migrating red knots is the coast of southern Brazil, but the final and most important pitstop is Delaware Bay, on the Atlantic coast of North America. During the last half of May, especially under full and new moons when the tides are at their highest, the sandy beaches of Delaware Bay attract one of the strangest of creatures, the horseshoe crab.

This marine animal is not actually a crab but a member of an ancient group of arthropods – related to spiders and scorpions – which fossil evidence suggests has remained relatively unchanged for more than 250 million years. Horseshoe crabs feed on marine worms and shellfish along the continental shelf, but in late spring they migrate inshore to spawn on protected sandy beaches such as those of Delaware Bay.

At night, or at dusk and dawn, masses of horseshoe crabs haul ashore. One or more males may be clasped to a female, attempting to fertilize the cluster of thousands of tiny green eggs she is here to lay. In one season a single female may produce around 80,000 eggs, most of which she buries out of reach of shorebirds. But waves and other horseshoe crabs can re-expose them, laying out a feast for the red knots and countless other migrating shorebirds on the Atlantic flyway. Eleven species of migratory birds, including ruddy turnstones, sanderlings and semipalmated sandpipers, also use the horseshoe crab eggs as a major food supply to replenish their reserves during a two- to three-week stopover. At peak crab nesting, the beaches of Delaware become a blizzard of wings.

On arrival here, a red knot, exhausted from its travels, might weigh as little as 90-120g (3-5oz), but on its departure in early June, it needs to have almost doubled in weight by laying down fat and protein reserves. This is both to fuel the remaining 2400km (1500-mile) journey to its Arctic breeding grounds and to ensure its survival on arrival at a time when food availability can be low. One estimate suggests that, to achieve this, the knot needs to eat as many as 400,000 horseshoe-crab eggs during its brief stopover.

Aerial surveys of red knots in Delaware Bay in the early 1990s estimated the population to be nearly 100,000 birds. But since then, numbers have declined steeply, to about 50,000 in 1999, and just 15,000 by 2008. Some people fear this race of red knot may become extinct within the next decade.

The decline is probably partly due to the loss of critical habitat, contamination and the spread of tourism at the knots' key wintering and migration sites. But the reduced availability of horseshoe crab eggs in Delaware Bay – as a result of increased harvesting of the crabs for bait in the conch- and eel-fishing industries – is very likely to be preventing the birds gaining the critical body-mass needed for their journey to the Arctic.

Though horseshoe-crab fishing is now being controlled and a sanctuary has been designated at the mouth of Delaware Bay, there are no signs yet of their recovery. Also, because horseshoe crabs do not breed until they are about ten years old, the situation remains precarious for the tiny red knot, whose breeding strategy is so dependent on the spawning of these prehistoric ocean wanderers.

Ostrich clutch control

Ostriches are the world's largest birds, with males up to 2.7 metres (8.8 feet) tall and weighing as much as 150kg (330 pounds). Their size renders them incapable of flight, but they have adapted to land life in remarkable ways. Though their wings are reduced to large plumes for display and temperature control, their long, muscular legs and two-toed feet can propel them at speeds of up to 70kph (43mph). And they have excellent vision from a high vantage point, with eyes that are 50cm (2 inches) in diameter. Another surprising thing about these avian goliaths is their nesting strategy.

Four subspecies of ostrich are found in subsaharan Africa: Masai, Somali, North African and South African. All live in semi-arid areas where there is a reliable supply of plant food and relatively open landscape, making it easier to spot potential predators such as lions and cheetahs. Studies in Kenya have revealed that when a male Masai ostrich comes into breeding condition, the pale pink skin on his neck turns bright red. He proclaims his territory with deep booming calls. He will mate with a number of females, often putting on a spectacular 'kantling' display, in which he squats and waves his wings alternately above his back as he rocks from side to side. Mating often happens near nest scrapes, which the male shows to the females. The first to lay in one of his scrapes is the one destined to guard and incubate – the 'major' hen. She lays 8-14 eggs there at two-day intervals, but surprisingly, she will also allow several other hens, often led there by the male, to lay in her nest.

Ostrich eggs are the largest in the world, weighing up to 1.9kg (4.2 pounds) with thick shells – though the smallest in relation to the size of the bird that lays it. Studies have shown that the 'minor' hens can add 3-20 extra eggs. But a major female can comfortably incubate only about 20 eggs, and somehow she knows which are hers and which belong to minor females, because almost invariably, it is the minor hens' eggs that she will push out of the nest – perhaps identifying them by their surface appearance, size or shape.

Opposite *Mating – the male's neck flushed red. He will mate with several females, close to his chosen egg-laying site. The first to lay in the scrape will become chief female, incubating the eggs of all his mates.*

Left *The male incubating the eggs at night. His dark plumage may be better night camouflage than that of the lead female, who usually takes the day shift. She is the one who has discarded some of the eggs – those in excess of the ones she can comfortably sit on, all belonging to other females who have laid in the nest.*

The major female incubates the eggs by day, sitting for up to 90 minutes at a time before moving position and sometimes turning the eggs. She must not only incubate them but also shade them from overheating. In dry desert landscapes, her brown colour helps camouflage her, and at night, the darker male incubates the eggs.

Throughout Africa, lions, spotted hyenas and jackals are major problems. Jackals can crack eggs together, and lions and hyenas have jaws powerful enough to break them. In East Africa the Egyptian vulture can crack the hard eggs by throwing stones at them with a sharp head flick.

Once the eggs have hatched, after six weeks' incubation, the dominant pair tend the chicks, constantly vigilant for predators such as martial eagles and hyenas. What's also remarkable is that they will sometimes allow or entice the chicks of other family groups to join their own – creating an ostrich super-crèche. It's thought this 'dilution' can help reduce the numbers of their own chicks lost to predators. The chicks that survive keep in a compact group and grow rapidly, reaching full ostrich height within a year.

The beetle-baiting owl

Above *The chicks at the burrow awaiting feeding time.*

Below *Father gathering dried cow or bison manure to lay at the burrow entrance.*

Opposite *A male at his well-manured burrow. While he's away foraging, his family can snack on dung beetles.*

Caledonian crows, Egyptian vultures and woodpecker finches are well-established tool users among birds, but in recent years scientists have added a newcomer to the list with a rather surprising 'tool use'.

The burrowing owl lives on open grasslands and agricultural plains across much of North, Central and South America. Apart from the larger snowy owl, which inhabits the Arctic tundra, it's the only one of its kind in the Americas that prefers to nest on the ground. And nothing could be more inviting than the burrows of the black-tailed prairie dogs that pepper the great plains of North America. Prairie dog 'towns' have numerous burrow entrances and tunnels as deep as 2 metres (6 feet) beneath the surface and up to 4.5m (15 feet) long. Abandoned tunnels are cool and reasonably safe hideouts for nesting burrowing owls.

Being relatively small creatures living on a big, open and potentially risky terrain, the owls and the prairie dogs need to be able to spot predators such as black-footed ferrets and ferruginous hawks from their burrow entrances. So the prairie dogs keep the surrounding grass well cropped, and the owls prefer vacant tunnels where the plants haven't grown too high. If there's a hint of impending danger, they are quick to alert one another – the prairie dogs bark and the owls give a distinctive *cak-cak-cak-cak* alarm call. It's good teamwork.

The burrowing owls in Conata Basin, bordering the Badlands National Park in South Dakota, arrive in early May after a winter spent to the south and begin courtship. This involves eye-contact, flashing white markings, cooing, bowing, scratching and nipping, as well as repeated display flights, when the male rises quickly to 30m (100 feet) and hovers for 5-10 seconds before plunging back down. Pairs will often nest in the burrow they used the previous year, lining it with dry material. Then the female lays her 6-9 eggs, which she incubates for about a month. The male is her provider, hunting mostly at dawn and dusk, bringing her a selection of food such as mice, grasshoppers, scorpions, frogs and small birds. And to top up the delivery service, he has a crafty ploy.

He will go in search of cattle or bison manure and other animal droppings, which he brings back in his talons and carefully places around the nest entrance and down in the burrow tunnel, stockpiling it. The dung acts as bait. Its scent attracts insects, including dung beetles, which may tumble into the nest. It's literally food on the doorstep. Studies in North Florida revealed this strategy to be so successful that, when dung is plentiful, it can provide a burrowing owl family with up to ten times their normal beetle intake. Some suggest that dung collection serves other purposes, such as acting as a visual sign to neighbours that the burrow is occupied, though this is yet to be confirmed.

Come late June, usually four or five fluffy chicks – sometimes more – emerge from the nest. Before long they are hopping about, tumbling back down their burrow and practising flapping over the prairie. As the chicks grow, their rasping food-begging calls are increasingly ignored by their parents, and when they are about six weeks old they leave the nest to try to catch their own insects. And when they find a mate, the likelihood is that some of the males will perform the cowpat trick that they inherited from their father.

A beak to hold a bird as big as its belly can

A wonderful bird is the pelican,
His bill will hold more than his belican.
He can take in his beak
Enough food for a week;
But I'm damned if I see how the helican.

DIXON LANIER MERRITT, 1910

Above *A great white pelican about to swallow a gannet chick and then feed the gutful to its own young. The Dassen Island pelicans also pillage gull, tern and cormorant colonies. It's a taste that has developed in the face of dwindling numbers of coastal fish.*

Right *Eyeing up another potential beakful – a gannet chick, not too large to swallow and with absent parents.*

Great white pelicans are among the largest flying birds, with a wingspan of about 3 metres (9.8 feet). They are also gregarious, breeding in colonies. Eighty per cent live in Africa, with most colonies on inland lakes but with an outpost about 9km (6 miles) off South Africa's Western Cape, on the small island of Dassen.

Great white pelicans began to settle and breed on Dassen in 1955, when about 20-30 pairs were displaced from Seal Island in False Bay because of a burgeoning Cape fur seal population and disturbance by harvesters of guano (seabird droppings). Today, Dassen's pelican population has multiplied to almost 700 pairs.

Great white pelicans can live for up to 30 years, reaching sexual maturity at the age of about 3 or 4. On Dassen, they begin nesting around August and usually lay two eggs, incubated for about a month. Normally just a single chick is raised, but even so,

Above *A chick of a great white pelican in the baking-hot colony. Its black coat may help in heat regulation. Though black heats up more quickly in the sun, the chick's black plumage doesn't let as much heat through as white fluffy down would.*

Right *The more normal, cooperative hunting strategy of pelicans – surrounding fish in a horseshoe formation and driving them into the shallows where they can be scooped up.*

finding sufficient food for it requires a lot of effort. Up until the 1970s, the Dassen pelicans would fly to the mainland to fish the freshwater wetlands and estuaries, sometimes hunting cooperatively, surrounding the fish in a horseshoe formation and driving them into the shallows. But in recent years, pelicans have found new food sources, eating chicken and pig offal and plundering rivers stocked with introduced fish. As a result, their numbers have increased significantly, but with the recent closure of some chicken-offal sites, they have switched their attention to seabirds.

At roughly the same time that pelicans nest on Dassen, tens of thousands of Cape gannets normally breed on the nearby island of Malgas. The seas are then at their most productive, and the gannets feed their ravenous chicks anchovies and sardines, which they catch by plunge-diving, sometimes from great heights. It's a spectacular event, with flocks of hundreds of gannets raining down into the sea like arrows.

In the past, while one parent was away fishing, the other tended to stay at the nest to look after the chick. But in recent years, with decreasing fish stocks, both gannet parents may be forced to go to sea, leaving their chick undefended. This hasn't gone unnoticed by the pelicans. Each day, hoards of them wander through the colony, hammering their beaks down on the chicks, filling their pouches and then swallowing their writhing prey. Only the bigger gannet chicks and those with a defensive parent on standby have a reasonable chance of surviving these lethal encounters.

As well as plundering gannet chicks, the great white pelicans have developed a taste for live chicks on their own nesting island, Dassen, feeding on colonial breeders such as Cape cormorants, kelp gulls, swift terns and even African penguins. This may seem gruesome, but with increased competition among the expanding pelican population and more hungry chicks to feed, it's not surprising they are taking advantage of their enormous bills to swallow easy prey.

Endurance parenting

Above *A chinstrap delivering krill to one of its fast-growing pair of chicks. To gather enough for the chicks to fledge before the Antarctic freeze returns requires a huge effort by both parents.*

Opposite *The chinstrap colony on the caldera rim at Baily Head on Deception Island. Volcanic heat makes it the ideal snow-free nest site for an early start to breeding. But the climb to the top is a test of endurance.*

The chinstrap penguin – named after the black line that circles its face from ear to ear like a helmet strap – is among the most pugnacious of penguins, particularly when it comes to chic- rearing. Breeding pairs are found on the Antarctic Peninsula and the subantarctic islands south of the Antarctic Convergence, the boundary where polar water meets temperate water. Deception Island, one of the most active volcanoes of the South Shetland Islands, is home to between 140,000 and 191,000 pairs. The largest colonies can be found at Baily Head, on the southwest flanks of the island, where an estimated 100,000 gather.

In October, as spring arrives, the first to return to the island head for the top, males vying for the best nest sites – free of snow due to the warming weather as well as the geothermally heated ground. Once a male has secured a site, he waits for his mate from previous years to arrive. After about five days, his attention may turn to another female, and if his original mate does eventually turn up, all hell breaks loose between the two females, the loser occasionally being tossed down

the hillside. On a roughly circular platform of small stones, two eggs are usually laid in late November, hatching in late December.

Chinstraps take it in turns to guard their chicks from aerial attacks from subantarctic skuas, while the other makes its daily foraging trip – a feat of endurance. Simply getting down to the sea is a challenge. The lava cliffs are steep and covered in ice, resulting in tripping and slipping. Then there are fast-flowing streams of meltwater to cross and blizzards obscuring precipitous cliff faces. And when the penguins finally reach the shoreline, they often face huge, pounding waves.

Chinstraps may swim 80km (50 miles) to feed, mainly on shrimp-like krill, and can dive to depths of up to 100m (328 feet). It's now that these flightless birds are in their element, using their wings to propel themselves at about 2m (7 feet) a second. After many hours at sea, they head home with a bellyful of krill for their chicks. Predatory leopard seals may be lurking to intercept the tired penguins. And then there's the marathon climb up to the rookery to find and feed their chicks, avoiding the pecks of neighbours and the harassment of skuas trying to make them regurgitate their food.

Reaching the nest, the tough decision is which offspring to feed, especially after the chicks have reached the age of 3-4 weeks and have mingled with others in dense aggregations – crèches. At this stage, a parent may react to its chicks' demands by running away. It could be that a parent is encouraging the chicks to explore the world beyond the crèche or luring them away from competing chicks from other families. Another possibility is that it is testing its offsprings' sprinting ability, discovering which is the hungriest by its determination to keep up the chase. These feeding chases may also be used to separate the siblings, so that the chicks don't compete with each other for a meal, reducing the amount dropped and wasted in a squabble. Whatever the case, the chick that runs the farthest tends to be the one that's fed – sustenance for a challenging life ahead.

The ultimate ice challenge

Chinstrap penguins may be one of the most numerous penguins in Antarctica, but their breeding distribution is limited by their preference for ice-free conditions for nesting. This explains why the biggest colonies, such as that on Deception Island, tend to be beside active volcanoes. On the volcanic island of Zavadovsky in the South Sandwich Islands, as many as 2 million chinstraps arrive in late spring to find nest sites. But farther south, where the snow remains into late October, colonies are smaller and the penguins have to race to breed and get the chicks to a size where they can take to sea before the autumn storms. The chinstraps breeding on rocky islands in the Rosenthal Archipelago form one such colony, overshadowed by the snow-blanketed glaciers of Anvers Island.

In late summer, leopard seals tend to hunt among the pack ice around the penguin colonies. This is the time the penguins are making daily foraging trips to keep their fast-growing chicks fed. But the adults are experienced at running the lethal gauntlet and have a good chance of avoiding attack. Not so their offspring.

In February, when the chicks are about nine weeks old, they lose the last of the soft, grey, downy feathers that have kept them warm these last two months, revealing the short, tough adult feathers that will form a protective shield against the freezing water and biting winds. At the end of the month, their parents abandon them. A chick will wait in vain for its parent to turn up on the shore. After a few days, driven by a desperate need to find food, it makes its way down to the sea. Gathered on the shore are similarly confused and hungry chicks, all flapping their flippers and slipping around on the rocks. Distracted by one another, they follow each other around on land for a while, but the urge to get in the water becomes stronger, and soon they are back on the seashore.

The first swim must be a shock. The water is icy, nearly two degrees below zero, and these chicks have never tried swimming before. But though they may

Above *Preparing for the plunge. The chinstraps have reason to be wary. To get to their offshore feeding grounds, they must run a gauntlet of leopard seals, which will pick off adults but especially inexperienced adolescents.*

Right *Taking the plunge. Chinstraps have the greatest chance of escaping predators when they swim under the water rather than scramble through broken ice. An experienced adult can outswim a leopard seal.*

Above *A leopard seal toying with a young chinstrap, making the most of easy pickings on the edge of the colony.*

Right *Young chinstraps about to take the plunge.*

Far right *A youngster attempting to struggle over brash ice rather than under it – a fatal move if a leopard seal is nearby.*

have a false start or two and retreat to the closest rock, it doesn't take long before they strike out away from the island. A hunting leopard seal does not even have to keep its head down – the chicks are utterly naive. They bob at the surface, flapping ineffectually, unsure of how to swim.

The Rosenthal Islands are close to Anvers Island, and the glaciers that slide down the slopes of Anvers form icy cliffs along its shores, which are constantly shedding pieces of ice, large and small. Fragments the size of tennis balls – brash ice – litter the water and are swept into bands by wind and tides. These bands of crackling ice move through the archipelago unpredictably, and when they bar the way to open water, it can be a catastrophe for a fledging chick.

Not yet knowing how to dive, a chinstrap chick enters a band of brash ice rather than turning back, attempting to push its way over the ice with its flippers. Its struggles attract the attention of a leopard seal cruising offshore. The chick battles on, skittering across the brash ice, unaware of the huge form swimming beneath it. A monstrous head rears up just behind the chick and then sinks again beneath the ice. The leopard seal takes its time. There is no need to hurry: the chick is helplessly stuck in the brash ice. Suddenly the chick vanishes from the surface, dragged down in the jaws of the seal. Under water, the leopard seal swims with its motionless but still live victim to an area clear of brash ice. It lets go and the chick starts to swim away, but the release is momentary. The seal grabs the chick in its front teeth and, with massive swings of its head, begins to flail it from side to side on the surface. Within minutes the chick is stripped of its meat, and the carcass is left to drift down to the seabed.

Of course, from each wave of youngsters launching from the beach, only a few will be taken, and most will reach the relative safety of open water. Here they will learn how to swim, dive and catch food, returning next season to the colony to start the whole process again.

Left *A leopard seal playing with its prey for just a few minutes before the kill.*

A tail of marvellous beauty

Opposite 'The hummingbird chased by a butterfly', his marvellous tail feathers trailing behind as he sips nectar.

Below The performer starting his show in a thicket but with enough light to make his iridescent feathers flash in the darkness. The finale is lift-off, when he pirouettes above the perch, his wings whirring and his tail feathers held high.

One of the rarest and most unusual looking of the 320 or so hummingbird species found in the Americas is the marvellous spatuletail. Found only in Peru, it is unique in having just four tail feathers. The outer pair are racket-shaped, and the male's ones are twice the length of his body, ending in disc-shaped 'spatules' that glint violet-blue. They can be moved independently, and in the breeding season, the male flaunts them in an extraordinary fashion, giving rise to its local name of *el colibri perseguido por una mariposa* – 'the hummingbird chased by a butterfly'.

Marvellous spatuletail hummingbirds are found in a handful of sites high up on the eastern forested slopes of the Rio Utcubamba valley in Cordillera del Colan. In the breeding season, October to May, male spatuletails congregate in thorny thickets at lekking sites – branches just a few metres off the ground used for displaying to passing females. Resting on his perch, a male resembles a small ping-pong ball, his tail-feathers draped below. But should a female appear, he raises his regal purple-blue crest and flicks his glorious tail above his head, sometimes twisting on his perch, his throat a glittering turquoise. And to ramp up the action, he takes to the air. Putting on a hover display, he pirouettes back and forth over the lekking branch seven or eight times, making a high-pitched *click* each time he flips over the branch. The flashing of his throat patch and crest in the dim understorey is presumably mesmerizing to

females – and possibly a sign of good health. The display lasts for 15-20 seconds, and should a rival male turn up, there may be a hovering showdown, with the pair 'facing off' until one retreats. Once the display is complete, the male wipes his beak on the branch, as if dusting himself off after such effort.

It takes lots of repeated displays at the lekking site to draw in a female. A male will return to the site every hour or so to try another dance, and if he does eventually succeed in luring a female onto the lekking branch, that doesn't guarantee she will mate with him. Recent evidence suggests that the males moult and regrow pristine tail feathers in time for the next breeding season. They are the only hummingbirds known to moult in this fashion – emphasizing the importance of their tails in attracting a mate and suggesting that, over time, selection by females has favoured the development of the fabulous butterfly tails.

Today, marvellous spatuletail hummingbirds are in a precarious position, decades of agricultural deforestation having reduced their population to less than a thousand. But their tails may yet save them. There is growing awareness of the hummingbird's beauty, and local pride in them, helped by education and increasing tourism. The children who once catapulted these glittering targets now sing songs about them and are proud to share their village with *el colibri*.

Time to sing and dance

When dead specimens of birds of paradise were first brought back to Europe by trading expeditions, they had no wings or feet. What the Europeans didn't realize was that these had been removed by native traders, in the traditional way, so the birds could be used as decorations. The Europeans thought that these were birds that never landed but merely floated in the forests, spirit-like, kept aloft by their extravagant plumes, and so they named them birds of paradise.

The rainforests of Indonesia, Torres Strait Islands, Papua New Guinea and eastern Australia are, indeed, paradises – so rich in food that feeding can be accomplished in a relatively short period, leaving time to devote to courtship. With rich areas supporting concentrated populations, competition for mates is rife, and courtship displays have become fantastically elaborate. Feathers have evolved into bizarre forms to impress females faced with so much choice.

Goldie's bird of paradise is found in forest on the mountain slopes of Fergusson and Normanby Islands to the southwest of Papua New Guinea. It gets its name from Andrew Goldie, who described it in 1882. The males display in groups of up to ten in a shared arena, or lek. They use a brilliant array of feathers to dazzle the watching females, but what distinguishes them are their diverse, if not particularly musical, vocalizations. When no females are present, males call to each other using a *wok-wok* or quiet *whick-whick* call. This becomes a loud, ringing *WHICK-WHICK* call if a female is attracted to the tree. As two plumed males begin to display, their calls begin with a ringing, metallic *WAAK* given first by one male and then the other but becoming more and more rapid until a continuous and drill-like sound reverberates through the forest.

As the male Goldie's duet, they perch facing each other with their bodies horizontal and their heads slightly dipped. Their wings are held open but low and are swept up and down as if rowing. The plumes are raised so that the shafts of the feathers are vertical and the long plumes spill downwards. The birds may also run up and down their display branches. The intensity of the display increases until one male stops performing and moves away from the centre of the display area to sit quietly and watch the rest of the proceedings. The victor stops moving about and slows down his rowing movements, making no sound at all. The comparatively dull female stands quietly near him for a while and then begins quivering her wings. Other, non-plumed, immature males, waiting on the sidelines, may mount her briefly, ignored by the victorious male. After a lengthy display, the male approaches the female, puts his neck and breast on her back and rubs to and fro. Then he mounts her, envelopes her with his wings and mates with her.

By comparison, king birds of paradise – the smallest of all the birds of paradise at only 16cm (6 inches) – have individual display area. They are found in the lowland forests of the Aru Islands, New Guinea and West Papua (formerly Irian Jaya). The male is a jewel-like crimson red with a white belly. His tail comprises two long tail-wires tipped by emerald-green discs.

Though diminutive, he has a large repertoire of calls, but these are mostly used to advertise his territory. The most characteristic call is a descending series of notes: *wher-whei-wha*. Sometimes as many as 15 notes are delivered in a row, mostly loud and fast but sometimes paced more slowly. Indeed the pitch and volume of the sequence can vary enormously. He also has a rising call, which is throaty by comparison, and another series of calls that can sound like the *miaow* of an angry cat. All these calls are about advertising presence, and so the females rarely call. But once the serious business of a display begins, the male's song becomes a continuous twittering and churring.

Male king birds of paradise spend almost all their time in their display territories, arriving first thing in the morning and leaving at five o'clock, day in day out. They find enough insects to satisfy their hunger on the

leaves and branches within their territories, together with fruit, leaving most of the day to 'make a song and dance' about their territories.

Males do not only display when females are present but, at various points in the day, they will display spontaneously. One sign that a male is about to perform is when he plucks a leaf or two from near one of his display branches. There are six phases to the display, not all of which may be performed in full. The wing-cupping display comes first, with the bird perched upright on his branch holding his wings partly open and vibrating them very fast. Next is the dancing display, which involves bringing his wings right around, close to his head, cocking his tail so steeply that the two tail wires bob around over his head, and then vibrating his body. If a female is watching, he performs with his back to her. Then he enters the tail-swinging phase, waggling his tail vigorously so the tail wires sweep from side to side over his head. This may be the conclusion of his display or he may enter the horizontal-open-wings display. With both wings spread forward, he vibrates them while rocking on top of his branch, and then he repeats the movements underneath the branch (the inverted phase of the open-wings display). For his finale, he closes his wings and swings pendulum-like upside down from his branch (the pendulum display).

If a female is present and is impressed, she will join him on the branch and quiver. He will rock to and fro, hitting her with his partly open beak. She then turns her back on him, and he mounts and mates with her briefly before she flies off into the forest.

This is probably one of the most exaggerated courtship displays in the animal world. It reveals how a tropical-paradise environment can free up species from the trials of food collection, making it possible for the males to invest energy in ever more fabulous feathers and dances, and for females to spend time being choosy about which males to select.

Left *A king bird of paradise on his display vine, ready to perform. He will dance several times a day, even if a female isn't watching, plucking a nearby leaf or two before he starts, as if getting himself in the mood. That he can spend all day showing off his skills is because his rainforest territory is so richly supplied with fruit and insects that he only needs to spend a short time feeding.*

The builder and decorator

Rather than investing in extravagant plumage to seduce females, the male Vogelkop bowerbird puts his efforts into crafting and maintaining an extraordinary roofed maypole bower – the most complex structure in the bird world. This work of art, bedecked with ornaments, is his show of health and fitness – a display site that he advertises with a cacophony of otherworldly territorial vocalizations, including whistling, rasping, gurgling, coughing, spitting and ratchetting. He will also mimic the songs of any birds in the vicinity, including parrots.

Only in the submontane and montane forests of the Arfak, Tamrau and Wandamen mountains of West Papua do Vogelkop bowerbirds construct these roofed bowers. The maypole is a column of sticks around a sapling trunk, and around that is woven a conical hut about a metre tall and 160cm wide (3.2 x 5.2 feet) with an arched entrance. The roof is typically made from orchid stems or sometimes sticks and ferns. The male covers the column base with moss, which extends out to form a huge green mat. This he decorates with piles of colourful fruits and flowers, beetles, butterfly wings, acorns and deer dung – the exact mix of which varies from bower to bower and between localities.

Such architectural craft requires constant rearrangement and redecoration with new treasures, as well as defence against raids by other males. There may be half a dozen other bower proprietors within a kilometre or so (half a mile) on the lookout for choice pieces to steal. A long mating season means intensive work for months on end, with males spending at least half their time perched in the vicinity of their bowers. Females tour and inspect each treasure trove, assessing their relative merits. A male responds to a female's arrival by hurrying to the back of his bower to hide, singing away. If he does manage to impress her, mating usually takes place on the edge of, or sometimes within, his prize-winning construction. Younger males tend to build inadequate structures with fewer decorations, and so only seasoned architects with big, colourful bowers get to mate.

chapter 7

Winning mammals

OVER THE PAST 65 MILLION YEARS, one group of animals has been so successful that it has defined its own geological era: the present-day Cenozoic, or 'age of the mammals'. Today, mammals comprise about 5000 species, half as many as birds – yet they dominate the Earth. If you doubt this, it's worth remembering that we humans are also mammals, and there are more than 6 billion of us and billions of our mammal livestock, pets and pests. We have reshaped the planet so that half of its productivity is directed towards our use. So how did we, the mammals, rise above all others? What has been the secret of our spectacular success?

The triumph of the mammals is such an unexpected story because we took so long to become important players. For most of our evolutionary history we have been small, secretive and unremarkable. We began our journey about 305 million years ago as primitive 'mammal-like' reptiles, but it took us an astonishing 100 million years to develop the characteristics we would recognize in today's mammals.

One of the earliest mammalian innovations was our unique feeding equipment. Unlike other animals with backbones, our lower jaw is just a single bone, studded with many differently shaped teeth. This new jaw and specialized dental tool kit (canines, incisors, molars, etc) allowed early mammals to bite and chew more accurately and forcefully and improved our ability to catch and process food.

The next major advance was to do with agility. Reptile legs sprawl out awkwardly to the side, forcing them to run by whipping their bodies back and forth. This sprawling method is fast when reptiles are small but has been improved on by mammals, which evolved limbs that lie close together under the body. This reduces stability but allows rapid change of direction when chasing prey or escaping.

By about 205 million years ago, the first 'true' mammals had arrived. They were shrew-like in size and behaviour, being nocturnal insect-eaters with small eyes and acute senses of smell and hearing. For millions of years mammals were forced by competition with large, day-living reptiles to remain small and avoid the day. But ironically, it was competition with dinosaurs and other reptiles that sowed the seeds of our future success.

Being nocturnal allowed mammals to develop excellent hearing and smell and the enlargement of those areas of the brain. These enhanced senses enabled sophisticated parental communication and led to the further enlargement of the brain and the uniquely mammalian neocortex region, which controls sensory perception, motor commands, spatial reasoning, conscious thought and language.

Being nocturnal also meant that early mammals had to develop ways of keeping their temperature high enough to remain active. They did this, like birds, by the

Left *A young male tiger – a modern mammal, possessing everything that makes mammals successful: a dental tool kit allowing specialized processing of food, acute senses leading to sophisticated communication, and a heating and cooling system enabling activity regardless of the weather.*

Above *A lion cub making use of
its mother's portable milk supply.
Mammals as a group are named
after the modified sweat glands
we use to provide milk.*

Opposite *A mountain gorilla
family. Lengthy parental care
and social interaction provide
the opportunity for youngsters
to learn from their elders.*

evolution of warm-blooded body chemistry, using food to generate heat and
retaining this warmth with insulating fur or fat. But to be warm-blooded, mammals
have had to increase their metabolic rate by as much as ten times that of reptiles,
meaning they have to eat ten times more. This is why mammals are always hungry.

The huge cost of food-gathering is partially recouped by having ten times the
aerobic endurance of reptiles to cover more ground to feed. Having a constant
body-temperature also led to the development of cooling mechanisms such as
sweat glands. These in turn led to a new way to provide portable food for infants
in the form of milk, which is produced from modified sweat glands.

Competition with reptiles had forced us to develop unique abilities, but 65.5
million years ago, mammals were still mostly insignificant creatures of the night.
Then something happened to change our fortunes. A large asteroid struck near the
Yucatan, in Mexico, plunging the Earth into cold darkness. In the ensuing
apocalypse, the larger day-living dinosaurs became extinct, giving the underdogs –
the warm-blooded, big-brained, nocturnal mammals – the upper hand.

After this great event, mammals were no longer constrained by dinosaur
competitors and flowered into the wonderful forms we know today – from the
minuscule bumblebee bat to the largest creature that has ever lived – the blue
whale. Crocodiles, lizards, snakes and the warm-blooded avian dinosaurs, the
birds, also survived, often to become deadly rivals of mammals.

By observing today's mammals, we can see many of the winning traits that helped
us to triumph. Polar bears display an amazing ability to thrive in extreme cold –
a lifestyle about as far as one can get from our humble beginnings. The strange
aye-aye illustrates the role that nocturnal senses played in our development and
how the ability of young mammals to learn from their elders fostered behavioural
adaptability unknown in other groups of animals – a primary reason for the
evolutionary success of mammals.

Sengis show the physical agility and endurance of mammals in comparison to
reptiles and thus how mammals may have conquered the day. Straw-coloured fruit
bats reveal the advantages of developing flight and the great feats of migration that
have resulted, as well as the beginnings of social coordination. Spotted hyenas give
us an insight into the evolution of complex societies that practise warfare, and the
competition between male humpback whales over mates illustrates how the
mammal way of life has allowed some to become the biggest and most spectacular
creatures to have ever lived. Together, such stories, told on the following pages,
illustrate mammalian beauty, character and adaptability that have helped us to
conquer the Earth.

The ice bear and the polar whale

No animal provides a more dramatic introduction to the story of the mammals' success than the world's largest land predator, the polar bear. It illustrates both our strengths and our vulnerabilities. But to understand the behaviour of today's polar bears, you need to travel back to the recent past.

About 200,000 years ago, on the tiny Admiralty Island group in southeast Alaska, a small population of brown bears became cut off by glacial advance. We suspect this because today's polar bears are more closely related to brown bears from these islands than all other brown bears are related to one another. In fact, one could argue that polar bears are really a type of white 'brown bear'. As these isolated brown bears found themselves increasingly surrounded by sea ice, they were forced to adapt to their frozen marine environment. Polar bears have fur and fat to conserve

heat created by their bodies, and they produce portable food, milk, for their young. These abilities have helped many mammals thrive in extreme polar regions where, for example, reptiles cannot.

As these early ice-living 'brown' bears began hunting seals on the sea ice, a number of other behavioural and physical changes evolved. Their teeth became better meat shearers than those of the other, largely plant-eating brown bears. White fur camouflaged their hunting efforts on ice, and a longer neck was better for reaching seals and for swimming long distances. To grip ice, their claws became shorter and stronger, and their feet developed knobbly traction pads. Crucially, they abandoned classic bear hibernation, as they could now hunt throughout the winter. This new 'polar' bear, the most recently evolved bear species, was so successful that it rapidly spread from the

Above *A potential opportunity of a bearded seal meal for a polar bear, made particularly difficult by the melting sea ice. Nearly all of a polar bear's normal diet is composed of seals. Rich in fat, they help fuel the bear's Arctic lifestyle.*

Opposite *A bear using brute strength and its enormous clawed paws to climb ice.*

Admiralty Islands across the Arctic. But in the process, it had become dependent on sea ice, from which to hunt ringed and bearded seals, which comprise its main food.

Today, Barter Island in the Beaufort Sea of northeastern Alaska has become one of the best places to observe polar bears. This little-known island is as remote and inhospitable as can be encountered in the United States. Here, the New World's western mountains become a flat coastal plain that meets the brutal Beaufort Sea. Each January, mother polar bears give birth to their young in snowy dens. They emerge in

March or April, normally with two cubs, and look for sea ice over the shallow (and thus productive) continental shelf where seals live.

As the summer progresses and the sea ice begins to retreat from the coast, polar bear mothers must make a monumental decision. Do they keep their cubs out on the melting ice, miles from land, or swim with their offspring to shore, where there is no food?

Scientists have been studying bear movements in the Beaufort Sea for many decades and are documenting the bears' amazingly complex decision-making and

recent challenges. In past decades mother bears could rely on the sea ice not melting beyond the shore-hugging continental shelf where the seals live. But as global warming has progressed over recent years, the ice often retreats more than 150km (93 miles) offshore in autumn. This has meant that polar bear families are increasingly choosing to risk the long swim to land, where they can conserve energy, rather than follow the distant pack ice over barren deep water.

Aerial surveys in 2004 discovered four dead adult polar bears that had drowned during a storm while attempting this crossing. Cubs are far less competent at swimming and so are particularly vulnerable and have suffered a 50 per cent drop in survival during recent decades. This is significant, because polar bears have one of the slowest reproductive rates of any mammal, with mothers typically having only five chances to give birth in a lifetime.

Once stranded on land, polar bears normally conserve energy by resting and living off their fat reserves. Fortunately for the bears of northern Alaska, the Beaufort Sea shore is a major migration route for bowhead whales. Bowheads often become stranded or fall prey to indigenous hunters, providing a life-saving source of food for hungry bear families while they wait for the sea ice to freeze again.

Polar bears have one of the best senses of smell in the animal kingdom, and experienced mothers know that this coast can provide a lifeline for them and their cubs. Barter Island is possibly the only place where you can to witness the rare social behaviour of these normally solitary animals and hosts the largest gathering of polar bears in the world, with up to 65 bears gathering to feed on whale carcasses. If a north wind blows, it is also the only place in the world where brown bears and polar bears can be found together again. This spectacular gathering of polar bears provides a reminder of both the remarkable adaptability of mammals and their fragility.

Right *A mother and cub gnawing at the deep-frozen food. This bowhead bonanza provides a lifeline for experienced mothers with cubs, gathering here in significant numbers.*

Opposite *A male taking a prime feeding spot. With food for all, there is comparatively little conflict between the much larger males and females with cubs.*

Learning to drum for a living

Madagascar's aye-aye is one of the most mysterious of living mammals. When it was first discovered in 1780, scientists thought it was a new species of squirrel because of its large bushy tail and continuously growing, rodent-like teeth. Later, though, its monkey-like skeleton revealed it to be the world's largest nocturnal primate and a close relative of the lemurs, which also evolved on the island of Madagascar. With its shaggy fur, large leathery ears, luminous eyes and long, thin fingers, it's also one of the world's strangest animals.

About the size of a domestic cat, the aye-aye creeps through the rainforest canopy at night searching for food. Madagascar has no woodpeckers, and so there is an ecological niche for an animal that can extract insects from wood, and this is the aye-aye's speciality. It drums its long fingers on branches and tree trunks up to 40 times a minute, listening intently with its sensitive ears. It can detect the subtle differences in pitch between solid wood and wood with small cavities created by the larvae of wood-boring insects. The aye-aye's hearing is so good that it might even be able to detect crawling grubs.

Once an aye-aye finds a cavity or insect, it uses its sharp front teeth to chew a hole in the wood away from the prey, which is trapped in a blind alley. Then it inserts its bizarrely thin, long middle finger and pulls out the grub. This finger has several special adaptations: it is curved, up to three times longer than the other fingers and extremely flexible, capable of moving 30 degrees sideways from its joint.

'Tap-foraging' is a complicated skill that takes a baby aye-aye years to learn. At birth, an aye-aye's ears are limp and only become controllable at about six weeks of age. Babies remain in the nest for one or two months but soon learn to climb on branches, hanging upside down. Gradually they become as agile in the trees as their parents, but even before they emerge from their nest, they begin to copy their mother's

tapping. As they watch her, they try to copy the delicate finger movements, spending a quarter of their active time practising.

If the mother locates food, the youngster rushes over and pushes her out of the way, greedily taking the prize. A young aye-aye eating a big grub resembles a human child eating an ice-cream cone. First it bites the head off and spits out the mouthparts – the portion that is hard to digest. The insect's insides then drip down the aye-aye's fingers. So it runs its tongue around its hand to lick up these juicy parts. But baby aye-ayes are also picky eaters, and they will wait for any new food to be approved by their mothers before they consume it.

Babies can't tap-forage until they're about 15 to 17 months old, and it takes perhaps two years to acquire the whole feeding pattern. Young aye-ayes become independent only at about four years of age, when tap-foraging produces 10-50 per cent of their diet. Interestingly, if aye-ayes are raised in captivity without an adult role-model, they fail to develop successful tap-foraging, suggesting that it is a skill that has to be learned.

Compared to its lemur relatives, the nocturnal aye-aye has an enormous brain for its body-size, probably because of the complex mechanics of its tap-foraging and its need for enhanced hearing and sense of smell.

The nocturnal and undeniably strange-looking aye-aye is often regarded by people in Madagascar as a harbinger of evil. Some believe that if one points its long middle finger at you, you are condemned to die, and the appearance of an aye-aye in a village is often taken as a sign that a villager will die, something that can only be prevented by killing the animal. Such superstitions have posed a threat to the species – which was once declared extinct and then rediscovered. A far greater threat today, though, is the steady loss of much of the aye-aye's forest home.

Above *A young aye-aye, with ears still floppy, practising the drum and finger-stab technique. It will have observed its mother's technique for several years.*

Opposite *An adult showing how to hook out a grub using its long, thin, flexible middle finger.*

Life in the fast lane

Above *A rufous sengi patrolling its trail network at great speed. The challenge is to find enough food to fuel such activity.*

One group of mammals has mystified scientists more than most – the sengis. When first described in the mid-nineteenth century, zoologists assumed these agile, bizarre-looking creatures were related to shrews. With their long, flexible, trunk-like noses and fondness for insects, their discoverers called them 'elephant shrews'. Subsequent generations of scientists attempted to reveal their true ancestry, deciding they must be distant relatives of antelopes, primates or even rabbits. Recent molecular evidence, though, indicates that they are part of a group of African mammals (the Afrotheria) that share a common ancestor and which include hyraxes, aardvarks, sea cows, tenrecs and, yes, elephants. In an ironic twist, it seems that 'elephant shrew' was more appropriate than the discoverers could have realized.

It has long been known that the rate at which mammals live their lives changes with size. Elephants, the largest living land mammals, do things slowly and are also the longest-lived. But the 15 species of the comparatively tiny sengis are forced to live life in the fast lane, surviving on an energetic knife's edge. The rufous sengi weighs only 50g (1.8oz) as an adult and lives in the dry shrubland of East Africa. With its 'go-faster' eye-strip, it seems like a bizarre cross between an antelope and an anteater. Much of its diet is of poor food value – termites and ants – and combined with its high metabolism due to its small size, the sengi's main problem is how to satisfy its hunger. Its solution is compromise and a bit of clever engineering.

Being constantly hungry forces sengis to be active through much of the day, which is fraught with danger, since movement attracts predators such as mongooses, birds of prey and reptiles. To outwit its enemies, the rufous sengi constructs a series of neatly cleared trails, which it memorizes in intimate detail, aided by regular scent-marking with its feet and tail. It patrols this trail system at breakneck speed, stopping to remove debris with a deft sideways sweep of its front feet. A single twig on a trail could have disastrous consequences, and so sengis spend 20-40 per cent of their active time running trails and removing obstacles. There is an added advantage to a trail system – like a mole tunnel, it helps make insects easier to spot.

Instead of making a nest or burrow, the rufous sengi acts like a small antelope, relying on above-ground bushy hiding spots. If it knows it's been spotted by a predator, before racing to safety, it often drums its rear feet on the ground, presumably to warn mates and young.

Sengis illustrate many of the advantages a mammal has over a reptile. With each long limb positioned under the body rather than to the side, sengis and most other mammals are more agile runners than reptiles of the same size. Being warm-blooded also gives mammals up to ten times the endurance of reptiles and makes them better at escaping. But reptiles have a few advantages of their own, and some smaller sengi species adopt a reptile tactic by reducing their body temperatures at night to as little as 5°C (41°F), going into a torpor that allows them to save 98 per cent of the energy they would consume if they kept their body temperatures high. Then they bask in the dawn sun to warm up again.

mummified. Resident crocodiles, hearing the breaking of branches, leave the water to subdue the unlucky.

Birds of prey are one of the mammals' most ancient and deadly enemies. Martial eagles, crowned eagles and African fish eagles survey the scene from emergent trees high above the swamp forest, as do many species of falcons, hawks, smaller eagles and vultures. But catching large bats is not as easy as it might seem. The birds of prey appear mesmerized by the scale of the bat colony and unable to decide which individuals to target. Sometimes they try to pluck roosting bats out of the trees, sending up great clouds of bats, or they attempt to attack them in flight, but their prey evades them by deftly dropping from the sky. Some bats are caught, but they are just a few out of the vast numbers. So the majority of bats live safely. The relatively low impact of birds on the bat colony illustrates another advantage of the mega-roost – predator swamping.

Each night for ten weeks, the bats consume more than twice their bodyweight in fruit, which means that the colony consumes 500 million kilograms of fruit during its stay – the equivalent of several billion bananas. Then, in just a few nights around Christmas, the entire colony leaves Kasanka. In terms of sheer numbers, this is the largest mammal migration on Earth. It seems amazing that this spectacle remained unknown to science until recently. But a larger mystery has remained: where exactly do the bats come from and where do they go?

Recently, Heidi Richter and colleagues placed satellite transmitters on four of Kasanka's fruit bats. The results were astounding. Each bat left Kasanka travelling north by a different route. One flew 370km (230 miles) in one night. One was tracked over weeks migrating more than 1900km (1180 miles) before disappearing deep in the Congo rainforest. This would make its round trip to Kasanka a minimum of 3800km (2360 miles) – one of the world's longest land-mammal migrations. Where the bats come from before they reach Kasanka is unknown. The mega-roost at Kasanka may have even

more important mysteries to reveal. Seventy per cent of fruit, nut and timber species used by humans in Africa are pollinated and dispersed by fruit bats. Could it be that the survival of the people and rainforests of Central Africa depends to a large extent on the survival of the bats in this one enormous bat roost?

There is only one known straw-coloured fruit bat colony in Zambia. But the evergreen swamp-forest which is its roosting habitat is fast disappearing and is now endangered. The privately run Kasanka Trust is working to protect the mega-roost, which will allow more people to witness this amazing sight and learn about the bats. What is heartening is that, even in our explored and exploited world, such great mammal spectacles can still be found.

Above *A straw-coloured fruit bat leaving the colony for the evening's foraging. Over ten or so weeks, it will consume more than twice its weight in fruit each night. Exactly where the bats come from and return to remains a mystery.*

Right *The great departing hordes. These fruit bats are vital to the forest's survival, and to that of humans – pollinating and dispersing the seeds of 70 per cent of fruit, nut and timber species used by Africans.*

Sengis are monogamous (have just one mate) and hold relatively large territories of 1600-4500 square metres (about an acre) that each defends from sengis of the same sex. This involves a spectacular display.

Rivals meeting on a territorial boundary walk slowly around each other lifting their long legs in a strut display, trying to look impressive. The ritual can suddenly dissolve into a blur of fur – a sengi fight that's over in seconds. A pair produces one or two precocial young – miniature replicas of adults, ready to go, with fur, sight and coordinated movement. Newborns are stashed in cover along the trail. The father provides no parental care, but he does help maintain the trail, defend the territory and warn of predators.

As small, nocturnal insect-eaters that hunt at night but which can also be active in the day, the high-speed sengis may give us an insight into how early mammals invaded the day.

Below *A rufous sengi foraging, mainly for termites and ants. It has an intimate knowledge of its trail system, which enables rapid escape. The sengi also keeps the trails clear of obstacles, both for ease of fast travel and to help it spot insects.*

The mega-roost mega-social

In 1986, a colourful British expat, David Lloyd, set out to explore a remote swamp in northern Zambia, only a few kilometres from the dangerous Congo border. Hearing rumours from local people of a giant bat colony hidden deep in the centre of the swamp, he set off to find it, making his way through the thick forest of twisted tree trunks and vines. In the distance, he began to hear a great clamour. It turned out to be the resonant screeching of millions of straw-coloured fruit bats. Lloyd had discovered for science one of the Earth's greatest animal spectacles.

Kasanka is flat, wet and covered with an impenetrable layer of rare 'mushitu' evergreen swamp-forest, and so at first, it was hard for Lloyd to comprehend the scale of the colony. Indeed, it was several years before scientists realized that he had found the world's largest fruit bat colony. Shortly after six o'clock each evening, between 8 and 11 million huge fruit bats take off from an area smaller than New York City's Central Park (0.5 square kilometres/0.19 square miles).

Bats are recent mammals, having first appeared about 50 million years ago, and had ancestors that were nocturnal tree-dwellers. Their wings have evolved from elongated finger bones and are covered by a delicate but fast-growing skin. The thumb of each hand is clawed and free of the wing, helping bats to clamber around in trees. Bats still have many limitations on their flight. They overheat when flying in direct sun and so tend to be active at night, which also helps them avoid competition with birds. Bat knees bend in the opposite direction to ours, allowing them to control their tail membrane in the air, like a rudder. However, their backwards-bending knees don't allow them to perch on branches like birds, and so they have to hang upside down when they rest, using special tendons to lock their legs into position when they sleep.

There are two groups of bats – the small, familiar, mostly insect-eating microchiropterans, which have tiny eyes and use echolocation to navigate, and the larger, tropical flying foxes, or fruit bats – the megachiropterans – which eat mostly fruit and nectar and find their way in the dark with large light-gathering eyes. Together, these two groups comprise more than 20 per cent of all mammal species and are a great mammalian success story. Taking to the air has allowed bats, like birds, to travel quickly and in an energy-efficient manner to exploit seasonal foods and climates, making them among the most numerous of mammals.

Straw-coloured fruit bats are among Africa's most widespread mammals, ranging from Mauritania in the north to the Cape in the south. In October, millions of these giant bats migrate from all over Central Africa to Kasanka in northern Zambia, forming the world's largest fruit bat roost. But the details of their journey and why they travel to this one small patch of forest have yet to be fully understood. Females arrive at Kasanka often in various stages of pregnancy, even though they rarely give birth there. This is a crucial clue – Kasanka is not a breeding colony. Elsewhere, colonies of this species give birth synchronously, suggesting that Kasanka's bats come from many and perhaps very distant groups.

To local Zambians, Kasanka means the 'place where people come to harvest', and it seems likely that that is why the bats arrive here at the beginning of the rainy season. From October through to late December, Kasanka supports an astounding abundance and range of fruit: waterberry, mupundu, loquat, figs and mangos, together with their leaves, pollen and nectar.

Every evening, 150,000 bats a minute leave the colony on long wings built for endurance, travelling up to 59km (37 miles) into the surrounding woodland to search for food. The bats are noisy eaters, and their large size restricts them to foraging in the upper layers of the canopy, which may prevent them from coming into conflict with farmers. And while they feed, they pollinate flowers and disperse millions of seeds of ecologically and economically important tree species.

At dawn the colony returns to the roost, illuminated by the rising sun and looking like millions of orange, butterflies circling the skies. With five times as many bats in one square kilometre than all the wildebeest in Africa, the colony is one of life's great spectacles. The bats are too big and too numerous to roost in tree cavities or caves and so huddle together on trees in the open. Leaves and branches are stripped bare by millions of claws and the sheer weight of bats. More bats arrive, filling the roost to bursting point. It is only when numbers of bats peak in November that the sheer scale of the roost becomes apparent. Only one other bat colony on Earth is larger, the well-known 20-million-strong aggregation of the relatively tiny Mexican free-tailed bats in Bracken Cave, Texas. These bats, though, are dwarfed by the giants of Kasanka.

With a wingspan of nearly a metre (80cm/30in), they form the world's densest concentration of mammal biomass. Perhaps 2500 tonnes of bats roost in this tiny patch of forest, equivalent in weight to 500 elephants. Yet few people have ever witnessed the spectacle.

In the crowded roost there is constant movement. The bats groom, nestle, sleep, flap and occasionally squabble. Overall, the impression is of a remarkable, harmonious existence. The constant chattering of the straw-coloured fruit bats is one of their many mysteries – we have no idea why they communicate so much or what they are saying. Periodically, whole branches and trees laden with bats are overcome by the immense weight and crash to the ground, and dead and dying bats litter the forest. Injured bats climb nearby trees only to become mummified. Resident crocodiles, hearing the breaking of branches, leave the water to subdue the unlucky.

Birds of prey are one of the mammals' most ancient and deadly enemies. Martial eagles, crowned eagles and African fish eagles survey the scene from emergent trees high above the swamp forest, as do many species of falcons, hawks, smaller eagles and vultures. But catching large bats is not as easy as it might seem. The birds of prey appear mesmerized by the scale of the bat colony and unable to decide which individuals to target. Sometimes they try to pluck roosting bats out of the trees, sending up great clouds of bats, or they attempt to attack them in flight, but their prey evades them by deftly dropping from the sky. Some bats are caught, but they are just a few out of the vast numbers. So the majority of bats live safely. The relatively low impact of birds on the bat colony illustrates another advantage of the mega-roost – predator swamping.

Each night for ten weeks, the bats consume more than twice their bodyweight in fruit, which means that the colony consumes 500 million kilograms of fruit during its stay – the equivalent of several billion bananas. Then, in just a few nights around Christmas, the entire colony leaves Kasanka. In terms of sheer numbers, this is the largest mammal migration on Earth. It seems amazing that this spectacle remained unknown to science until recently. But a larger mystery has remained: where exactly do the bats come from and where do they go?

Recently, Heidi Richter and colleagues placed satellite transmitters on four of Kasanka's fruit bats. The results were astounding. Each bat left Kasanka travelling north by a different route. One flew 370km (230 miles) in one night. One was tracked over weeks migrating more than 1900km (1180 miles) before disappearing deep in the Congo rainforest. This would make its round trip to Kasanka a minimum of 3800km (2360 miles) – one of the world's longest land-mammal migrations. Where the bats come from before they reach Kasanka is unknown. The mega-roost at Kasanka may have even

Above A straw-coloured fruit bat leaving the colony for the evening's foraging. Over ten or so weeks, it will consume more than twice its weight in fruit each night. Exactly where the bats come from and return to remains a mystery.

more important mysteries to reveal. Seventy per cent of fruit, nut and timber species used by humans in Africa are pollinated and dispersed by fruit bats. Could it be that the survival of the people and rainforests of Central Africa depends to a large extent on the survival of the bats in this one enormous bat roost?

There is only one known straw-coloured fruit bat colony in Zambia. But the evergreen swamp-forest which is its roosting habitat is fast disappearing and is now endangered. The privately run Kasanka Trust is working to protect the mega-roost, which will allow more people to witness this amazing sight and learn about the bats. What is heartening is that, even in our explored and exploited world, such great mammal spectacles can still be found.

Right *The great departing hordes. These fruit bats are vital to the forest's survival, and to that of humans – pollinating and dispersing the seeds of 70 per cent of fruit, nut and timber species used by Africans.*

The heavyweight contest

One of the most remarkable mammal spectacles is the fin-slapping, chin-slamming, breaching, bubble-blowing, jousting contest that takes place among male humpback whales. Described by biologists as the 'heat run', this spectacular competition involves as many as 40 males struggling, often violently, for position beside a female.

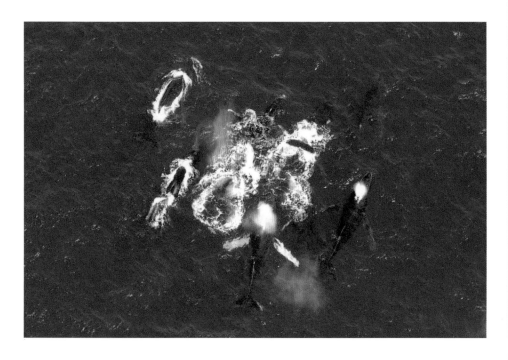

Above *Above-water signs of a below-water contest involving up to 40 male humpbacks competing for females.*

Right *An escalating fight. Rival males may battle for hours, often charging at each other, which can result in injuries and sometimes death.*

The humpback is probably the largest animal in the world to fight over females. When fully grown, a male averages 15.5m (50 feet) long and weighs up to 40 tonnes. (Females are even larger, weighing 44 tonnes or more.) The mating contests take place in tropical waters, where the humpbacks migrate to each winter from their polar feeding grounds – a journey of up to 4000km (2485 miles). There is, though, little food for them in the tropics, and once there, they are forced to live off their fat reserves. So why don't they just breed at their feeding grounds? Pregnant females probably make this journey because the higher temperatures help their calves survive their first weeks of life, and the likelihood is that it's their choice of calving

grounds that has determined where mating takes place with the males just following their lead.

The main challenge for a male humpback is how to find a potential mate in the huge but barren tropical oceans. One solution is song – a complex vocalization, which seems to have much in common with bird song and contains powerful, low-frequency elements that could well carry hundreds of kilometres through the water. The breeding grounds vibrate with a cacophony of song as the males hang suspended deep under water, repeating parts of their 10- to 20-minute songs for hours, often at night. It's uncertain whether the songs are directed at females, rival males or both, but they change and evolve from year to year in a sort of whale 'pop chart'.

The second challenge for a male is to find a female who is receptive, given that females come into heat for possibly only one or two days. It seems likely that females advertise when they are receptive by releasing a chemical scent into the water. Interestingly, males have been observed surfacing with open mouths, apparently tasting this scent in the water.

As males begin to gather around a female, she starts to swim away, and the pack follows her at speed. One of the larger males, the 'primary escort', positions himself directly behind the female, while smaller and often immature males hover on the periphery, possibly learning to compete. If there are any large males of similar size to the escort, then a contest takes place for pole position. The warning shots are fin-slapping,

breaching and bubble-blowing. Bubble streams may act as a visual barrier, screening the female or threatening rival males. As things escalate, chin-slamming starts, the males following each other and repeatedly rising out of the water and slamming their lower jaws loudly on the surface. If the contest intensifies, males will charge each other and attempt to push each other under water. Competing males may throw themselves at one another or leap out of the water on top of each other with such violence that physical injuries occur. Males may even be killed in these dramatic battles.

Human divers trying to film or study the heat run experience groups of truck-sized whales passing over them in pursuit of the female at great speed – an experience that apparently feels like driving the wrong way on the motorway. But only from the air can the full scale and power of the heat run be observed. These contests may continue above and below the surface for hours. As the escorts and challengers fall by the wayside, a victor eventually emerges to swim with the female. But the next part of the story remains a mystery. Despite the many thousands of hours that scientists have spent observing them, no one knows where the whales mate, though it seems plausible that it takes place in the depths.

Why do these enormous whales go in for such exhausting and ultimately dangerous contests? The likelihood is that these great tests of stamina are an efficient way for females to evaluate male fitness quickly and so find the best possible mate in the vast ocean.

Above *A competing male blowing bubbles, possibly to screen off his female from rivals or to intimidate them. Where they mate is still unknown.*

Next page *Breaching giants, displaying their size and strength with massive crashes.*

The female warrior society

For thousands of years, spotted, or 'laughing', hyenas have spawned disturbing myths about their deviant and demonic behaviour. Humans have had a long association with spotted hyenas, first in Africa and later in ice-age Britain and Europe, where they are now extinct. But only recently have scientists begun to understand their biology and reveal facts more bizarre than the unsettling legends ever were.

Despite their dog-like appearance, hyenas are more closely related to cats, mongooses and civets. Of the four living species, spotted hyenas are the largest and most unusual. The females look and act like males, with a resemblance that includes their genitals. Their reproductive system is unique among mammals – with a long, tube-shaped pseudo-penis, the size and shape of the male's penis and formed from the clitoris, complemented by a pseudo-scrotum. Together these mimic male genitals, and females must urinate, mate and even give birth through this pseudo-penis. Because of this strange anatomy, birth is extremely dangerous, as the first-born must rupture the female's pseudo-penis. Three quarters of first-born cubs die during birth, as

do 10 per cent of first-time mothers. A quarter of mammal families have species in which females are larger than males, and several have females whose genitals mimic males' (including spider monkeys, lemurs and European moles), though none as successfully as the spotted hyena.

So what is the advantage of being a butch female? One theory is that it's a consequence of the species' highly competitive communal feeding. Females are more masculine than males – larger, more aggressive and dominant over males. The most intimidating females and their young eat first, and that favours increased aggression and high levels of male hormones. But competition for food can't be the only reason behind the female spotted hyena's male-like character, as the females of many other carnivores compete over food.

Hyena society is about managing aggression and cooperation – and it starts at birth. Though females normally have twins, after a few weeks, up to a half of litters are lone cubs. Spotted hyena cubs are unique among mammals in being born with functional fangs,

Below *A losing battle between a lone spotted hyena and its deadly enemy, the lion. For hyenas, safety comes in numbers.*

and soon after birth, the babies begin fighting violently, one often killing the other. Fighting is particularly deadly between same-sex twins. The narrow aardvark burrows in which they are born protect cubs from lions but not from each other and are too small for mothers to enter and stop fights.

Hyenas are excellent mothers, but they face many challenges. They often need to travel great distances to find food, and cubs may be left on their own longer than any other baby mammals – up to a week. A mother may make up to 50 hunting trips – a possible 3600km (2237 miles) a year – returning to feed her young on rich milk. Mothers also help cubs develop social skills quickly so they can integrate into their complex clan society of between 3 and 80 members. Spotted hyenas are thought to be highly intelligent and have a remarkably sophisticated social system.

For female cubs, social standing is inherited from their mothers, and the offspring of alpha females have feeding priority within the clan. One of the advantages of this complex society is cooperative hunting. It's a myth that spotted hyenas are scavengers: they are among Africa's most skilful predators, hunting up to 70 per cent of their own food. Clans split up into smaller hunting parties and target prey as big as zebras, wildebeest and even Cape buffaloes, giraffes and young elephants. They are endurance athletes, with large hearts

that allow them to trot at 10kmph (6mph) for great distances without tiring and to chase prey at speeds of 50kph (30mph) for more than 3km (nearly 2 miles). Some chases may continue for up to 24km (15 miles).

Hyena clan structure is useful not only for competition with other clans, but also against their deadly enemies – lions. Different species normally ignore each others' territorial boundaries, but hyenas and lions defend boundaries against each other as they would against members of their own species. Lions will kill hyenas whenever possible. Conversely, hyenas are major predators of lion cubs and, if they have sufficient numbers, can even kill adult lions.

A hyena clan's success at defending itself from lions or appropriating their kills is determined by the presence of male lions and whether the clan can recruit enough members to do battle. Encounters can be deadly. One battle between lions and hyenas in Ethiopia's Gobele Desert escalated into a 'war' that lasted two weeks. The lions eventually won, driving the hyenas away after killing 35 and losing 6 of their own. Nevertheless, the spotted hyena remains Africa's most common and successful large predator. Their complex social behaviour provides important insight into how adaptable mammal societies have evolved and succeeded, and this may help us to understand the origins of societies and even warfare.

Above *A large lion pride, refusing to relinquish its catch. Once the hyenas recruit enough of the clan to help, they are capable of taking over the meal.*

Right *A hyena clan combining to drive lions out of its territory. Strength in numbers and cooperation are what counts in both societies.*

chapter **8**

Hot-blooded hunters

OUR ABILITY TO LEARN is one of the main reasons for the success of us mammals. If you can make use of knowledge gained from your successes and failures, you can rapidly adapt to the problems thrown up by a particular habitat or a changing environment. Most mammals, especially longer-lived ones, also invest time in caring for their young. Youngsters who learn from their parents benefit from the knowledge gained by them over their lifetimes, and so in turn inherit a huge advantage over their rivals. The adaptability that results from the capacity to learn is what has enabled mammals to exploit some of the harshest of environments.

This chapter focuses on the adaptable behaviour of individuals or populations of mammals when it comes to hunting prey or avoiding becoming prey. Sometimes this involves developing strategies unique to a situation or location. Take the cheetah brothers of Lewa Downs. It's possible that these are the only cheetahs in the world to routinely hunt ostrich. They do so not because they have to but because they have learnt how to. The ability to catch such a potentially dangerous animal through a cooperative hunting strategy has given the brothers an advantage over other cheetahs in the area and has allowed them to hold their territory for nearly a decade.

But for all predators, however developed their strategies, conditions must be at their best for hunting to be successful. This might mean optimal weather. Rain can ruin a bat's chances of hunting, and in the case of the fish-catching bulldog bat of Belize, even wind rippling the water surface can foil a hunt. So a hunter must be ready to exploit every opportunity. In some cases, this requires building up detailed knowledge of a territory. A female killer whale who hunts around the Falklands has learnt to exploit a food resource that may only be available to her for a few days a year. And because her calf follows her into the hunting zone, the knowledge of how to pull off such a dangerous manoeuvre is being passed down from one generation to the next.

But, of course, it is not just the hunters who are adaptable. Prey species have numerous strategies to avoid being caught. Just as a hunter's actions are focused on the moment when it can strike, so a prey animal must never drop its vigilance, even for a moment – because that's the moment the hunter will have been waiting for. So the lives of predator and prey are intertwined. In cases where a predator is a specialist on one type of prey, this means that the fortunes of the populations themselves are linked. Instinct clearly plays a huge part in the survival of individuals, but experience may even be passed down to future generations through inheritance, as in the case of the snowshoe hare, where hormones play a surprising part in regulating the balance of numbers of both predator and prey.

The natural world is full of drama and spectacle. But the moment when a perfectly adapted predator pitches its skills against those of its equally highly tuned prey are the moments that so often fascinate us and which encapsulate the struggle that is life.

Left *A lioness charging across a river in the Okavango. She is in pursuit of her pride, who are calling to her as they start trailing a herd of buffalo. The success of the hunt depends on the pride members all playing their part as a team.*

Previous page *A killer whale ramming a grey whale calf off the coast of California. It is one of a group of transient whales who specialize in hunting marine mammals. They have learnt when and where to expect the grey whales and their calves on their migration up the coast. Though the mother grey whale will try to protect her calf, she is no match for the cooperative hunting skills of the killer whales.*

Fastest cats versus biggest birds

Above *The Lewa Downs band of brothers. It's a coalition that has allowed them to prey on large animals, even ostriches, which single cheetahs can't.*

Right *One of the brothers stalking prey. The cheetah hunting strategy relies on a burst of speed rather than stamina, but as a team, the brothers can work in relay, one making the first sprint, the second taking over as it tires and the third making the final burst.*

Many mammalian hunters live in groups – wolves in packs, lions in prides, orcas in pods. There are advantages to this lifestyle for holding a territory, hunting and the care of young. Usually groups are mixed-sex associations, but for cheetahs, group living most commonly involves coalitions of males, often litter brothers.

Seven or eight years ago, three fully grown male cheetahs appeared together on Lewa Downs in the north of Kenya. Physically, two were similar, while the third was of a slightly lighter build but similar enough to be presumed a brother. When the trio first arrived, they were nervous and hard to observe, having come from the unprotected areas to the north. As time has passed, they have become more tolerant of people, and as it has become possible to observe them, their fascinating hunting strategy has begun to be revealed.

This region is dry country – a mix of rocky hills and rolling open plains that stretch to the mountains on the northern horizon. It's inhabited by a diversity of species, many adapted to cope with unpredictable rainfall.

Cheetahs don't have the mass of a lion or leopard and don't normally try to pull down and subdue large prey. Instead, they rely completely on speed to catch their food. So any injury which might slow them down could soon prove fatal. But the brothers don't follow the rules and frequently take on extremely large and dangerous animals.

It might be thought that a zebra has little defence other than running away, but this is far from the truth. Zebras are dangerous at both ends, capable of delivering serious bites and stunning kicks. A fully grown zebra can be lethal to a lion, let alone a cheetah. But despite the risks, the brothers often hunt zebras. The strategy is straightforward: stalk in close enough to start the run, try to isolate a foal or a small zebra and pull it down, either by landing with a thump on its hindquarters or by knocking its legs out from under

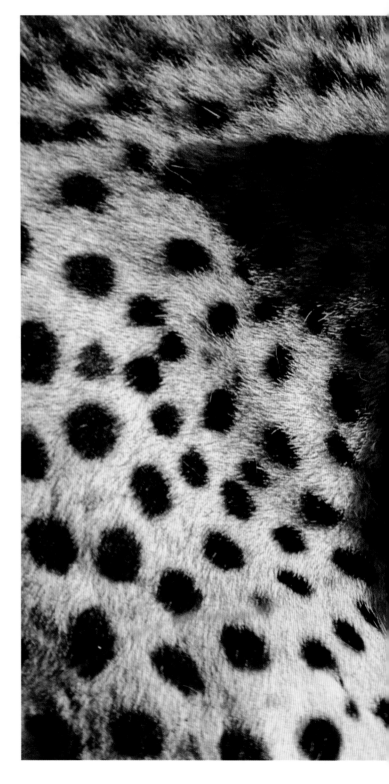

Above, left to right The start of the hunt. As a male ostrich walks by, it's spotted by one of the resting brothers. The cheetah starts the stalk – and then the sprint. The female ostrich joins her fleeing mate, only to become the target. The second brother now charges in, hooking onto her. The third brother joins in, and the combined weight and power of all three bring down the huge bird.

Opposite The feast. The brothers feed intently, keeping watch for lions and hyenas, which could easily appropriate the meal.

Overleaf The Lewa brothers scanning for prey. Cheetahs are normally solitary hunters, but these male siblings have learnt to work as a unit when hunting.

it. But the execution is far more perilous – female zebras are tenacious in the defence of their young, and even the herd stallion may join in the defence with lashing hooves and bared teeth.

It often requires the combined efforts of the three brothers to subdue their prey, and many hunts take the form of a relay – one cat taking up the initial running, the second taking its place, followed by the third, which might finally be able to knock the zebra off its feet, to be joined by his brothers for the finish.

They've also been seen deploying the same strategy with other, even better-armed prey. But often even the brothers' combined efforts aren't enough. An oryx with a broken front leg defended itself with the threat of its rapier horns for two hours before the brothers gave up. A young eland would have been easy to pull down had it not been for the concerted and extremely aggressive response of the adult elands – the largest antelopes in Africa – which put the brothers to flight.

The three cheetahs appear very large by normal standards, and with the confidence of strength in numbers, they are able to have a go at the most unlikely prey – even adult ostriches. On average, the brothers take one ostrich a month. Each hunt is different. The brothers might come across an ostrich while patrolling their territory or when they are

actively hunting. But just as frequently, an ostrich unwittingly approaches the tree under which the brothers are dozing, heads on chests.

On seeing the ostrich, they come to their feet with a slow, easy grace, taking extreme care not to alert it to their presence. As one cat takes the lead, the other two hang back, waiting for a cue to act. The lead cat's head drops slightly lower than his shoulders as he glides forwards – eyes glued on the target. Each step is weighted and considered, his huge shoulder blades rising and falling as he ghosts forward, exuding a sense of pent-up energy.

Alternately edging forward, freezing or dropping to the ground, the cheetah starts to cover the distance between himself and the ostrich. Each time the ostrich raises its head, he freezes. Meanwhile, the other two start to follow 30 or so metres (98 feet) behind. Keeping tabs on all three is almost impossible in Lewa's long grass, especially as the cheetahs use every crease and fold in the ground. So skilled are the cats at stalking that even the advantages of the ostrich's long neck and sharp eyes are outweighed.

By now the lead cat has closed to within 40 metres (130 feet) and vanished into a patch of grass. Suddenly the ostrich lurches violently away and starts to run at full tilt. From a few metres beyond the place the

cheetah was last seen, a blur of movement erupts. As the ostrich runs, its long legs rake forward, and it accelerates so fast that catching it would seem almost impossible.

But the cheetah is racing, too, eyes locked onto the bird. Suddenly it accelerates, closing in and displaying why it's the fastest runner on the planet. Now it seems to be flowing over the ground, feet barely touching the surface, racing directly behind the ostrich. Launching itself onto the bird and hooking on with his front feet, the cheetah desperately tries to drag the ostrich back, but the bird is travelling so fast that it carries the cat with it. But the cheetah's back legs are hitting the ground, and this and the weight of the cat start to slow the bird down. Now the second cheetah comes charging in and grabs the ostrich by one wing, spinning the bird off its feet. The third cheetah then latches onto the bird's throat just below the head, pulling it back and away from the body to avoid the ostrich's flailing feet, which could eviscerate the cheetah should they connect.

Suddenly it's all over. The cats have to be quick about feeding, because Lewa is home to lions and hyenas, which could easily drive the brothers away. They feed with none of the squabbling associated with the other cats – and at all times, one will have his head up, checking for danger. Perhaps because of the body

shape of the ostrich carcass, the brothers seem unable to turn the bird over to eat the second half. But by then, the brothers' bellies have grown to an incredible size, and it will take a few days of lying under a tree to digest the meal.

Ups and downs of the harried hares

Right *A snowshoe hare nibbling buds. It was once thought that the cyclic population crashes of hares were caused by a lack of food, the result of overgrazing. But the trigger turns out to be the number of predators that start to feed on a plentiful hare population.*

Opposite *A predator that specializes in feeding on snowshoe hares – the Canadian lynx. Its own reproductive success is tied to that of the hare.*

There are few wilder places than the Yukon. Winters are long, cold and harsh. Temperatures frequently drop to 40 below, and when the wind picks up, exposed skin freezes in seconds. Deep snow and the rugged terrain makes any journey on foot an enormous effort. But winter is also a time of great beauty. Snow softens the lines of the landscape, lending the mountains and forests a fairy-tale aspect, and the northern lights fill the night skies with ever-changing curtains of light.

In the calm of a crisp morning, animal movements can be read in the snow. Footsteps tell of wolf packs travelling great distances, snowshoe hares hopping from willow patch to willow patch or wolverines lumbering along checking out every possibility of food. Not surprisingly, many of the mammals have luxuriant fur coats, and the money that could be made from these furs attracted trappers here more than 300 years ago. Along with the pursuit of gold, the trade in furs

led to the opening up of this incredible wilderness. The Hudson Bay Company was central to the fur trade in Canada, and it kept careful records of the numbers taken each year. In the early 1930s, analysis of these records led to a fascinating revelation.

It became apparent that, over a period of eight to eleven years, both lynx and snowshoe hare populations would go up and down in synchrony with each other. It seemed that the snowshoe hares would be the first to hit a peak and then suddenly go into a rapid decline, a decline that was soon mirrored in the lynx population. Over a number of years, both populations would continue to decline. After having bottomed out and remained at this low level for a number of years, they would then slowly increase again, achieving another peak roughly ten years after the last. What this told scientists was that somehow the lynx and snowshoe hare populations were directly connected to each other and that the decline of the hare provoked the decline of the lynx. But what made the hare population crash?

It used to be thought that the hares' numbers may have got so high that they literally ate themselves out of house and home. It is true that this population does get very high, with as many as four hares per hectare at the peak, but recent research has suggested that the decline's main cause is predation. With such a glut of food available, many predators focus totally on the hares. Lynx, fox, wolf, coyote and wolverine all hunt snowshoe hares, as do owls and other birds of prey, and when the hare population is at its peak, other less expected predators get in on the act. Kestrels and even red squirrels will take whole litters of young hares. So intense is this predation that more hares are being eaten than are replaced through breeding.

As the snowshoe hare population goes into a rapid decline and there are far fewer hares about, many of the predators will starve, and the ones that survive have to switch to other prey. But for the lynx, switching

Right *A race for life, for both predator and prey. The stress that increased predator numbers puts on the hares causes females to produce fewer young. They in turn inherit high levels of stress hormones, which makes them even more wary.*

isn't an option. It's so specialized in hunting hares that its fortunes are totally tied to those of its prey, and so in turn, its population goes into rapid decline. But what has not been clear is why the hare population remains at such a low level for so long. One would think that an animal that can reproduce at such a frantic rate would soon bounce back once the predator pressure was off. The most recent research suggests a fascinating answer to this conundrum.

When the hare population hits a peak, the predator pressure on the hares becomes intense, and the hares become increasingly stressed. In the case of females, this causes them to have fewer litters and fewer young in each litter. So just before the population goes into rapid decline, more hares are being killed by predators exactly when far fewer young are being born. No wonder the population crashes. But what researchers are suggesting is that the daughters of stressed females are born stressed. It appears that the stress hormones present in the mother affect her unborn young, who in turn also have fewer litters and fewer young. It's almost like a stress hangover that lingers generation after generation, all caused by the earlier massive predation. It is only some three to five years after the intense predation that the hares finally start to breed normally again, at which point the population starts to increase.

Can this inherited stress be of any benefit? On an individual level, it could be a huge bonus. The young of a stressed female might well be much more alert and wary and so safer, because they would be much more likely to spot predators. In a situation where there are large numbers of predators still focusing on the hares, it would be far better to have fewer, more wary young than to have lots of naïve young. It's only when the predator numbers have dropped that it makes sense to start having more offspring that are less wary.

The lives of predator and prey are interlinked, but in the case of the lynx and the snowshoe hare, the relationship is one of the most extreme in the natural world.

The night-flying fish-detectors

Above *The great muzzle of a bulldog bat, its cheek pouches stuffed with fish.*

Opposite *Bulldog bats leaving their daytime roost in the trunk of a rainforest tree, heading for the river. It's a flying mammal that can fish in the dark.*

All mammalian hunters push the boundaries. Unique adaptations of behaviour and finely tuned senses allow them to exploit food that remains out of reach to other, less adaptable groups. But success involves more than just the right tools – precision timing is required to take advantage of windows of opportunity.

Belize is a land shaped by water. Over the millennia, it has eroded the limestone hills, and divided them with clear-water rivers. The rain that falls so frequently here fashions and nurtures all elements of life, especially the rainforests. One group in particular does well here – the bats. Many use echolocation to locate prey. They literally shout into the darkness and then listen for the echoes, from which they build up a picture of what's ahead. But the greater bulldog bat has a bigger problem than just seeing in the dark – its prey lives under water.

Bulldog bats are quite large, and the distinctive muzzle that gives them their name lends them an air of menace. Each evening they emerge from their roosts and head to the rivers. They fly in almost total silence a foot or two above the water's surface, with a distinctive stiff-winged, choppy flight, zig-zagging back and forth looking for ripples – ripples that might betray the presence, just beneath the surface, of a shoal of small fish. At the merest hint of movement, a bat will drop down and, as it passes over the spot where it thinks the fish are, rake its feet through the water.

The bats' feet are key to the success of this strategy. Each foot is thin at the ankle but broadens through the length of the foot to the long toes, which are armed with hooked claws. The foot is dragged through the surface of the water with the soles facing forwards so the claws hook forwards like grappling irons. Any fish near the surface will be snagged by these fearsome weapons. Often a single claw will slice into a gill cover, firmly hooking the fish, which is then dragged from the water and carried forward by the bat's momentum. The bat continues its forward flight, but it now has a wriggling, writhing fish in its grip. Beating its wings

powerfully, it starts to climb, and as it does so, it swings the fish forwards and up, while at the same time bending its head down and hunching right over. The fish is transferred to the bat's mouth where it is dispatched with a bite to the head. The bat will then happily eat the fish while still on the wing – stuffing meat into its cheek pouches, which soon bulge with the spoils of its success.

What's extraordinary about this behaviour is the dexterity and coordination the bats show. They are capable of great speed, and to hit a fish while flying at more than 64kph (40mph) so close to the water's surface requires huge skill. But the bats have more than just this one strategy.

If there is no sign of fish at the surface, then the bat uses its fantastic spatial memory to return to the spot where it last successfully launched an attack. The bat glides close to the surface and lowers its legs into the water as before. But this time, the bat drags its feet through the surface for up to a metre (3 feet). Gliding along just above the water, the bat looks like an ice skater, the force of its feet making a distinctive *fruuuuuup*. These speculative rakes are remarkably successful – it might be that the speed of the bats coming through the hotspot means that the fish literally never see them coming.

The bats will fish like this with peaks of activity throughout the night. But there is competition for the best fishing spots. The fish are only at the surface for a brief time, and the bats have to exploit this while they can. On small pools, there is a risk of collision, and here once again echolocation helps. If one bat is on a collision course with another, it will 'honk' at it by dropping its echolocation call an octave. The other bat, on hearing this, will usually give way.

How did this remarkable strategy come about? It may well be that the bats started by taking insects from the surface, and catching fish was a step on from there. In any event, it is a unique combination of skills and abilities that has enabled the bats to become fishers.

Above *The catch. Enormous
claws – proportionally larger
than those of a tiger – are used to
hook fish at the water's surface.*

Left *The action. A bulldog bat
sweeps the water with a pulse of
sound, listening for the echoes of
any surface movement. It locates
potential prey, lowers its
grappling irons and hooks a fish,
dragging it out of the water with
the momentum of its flight.*

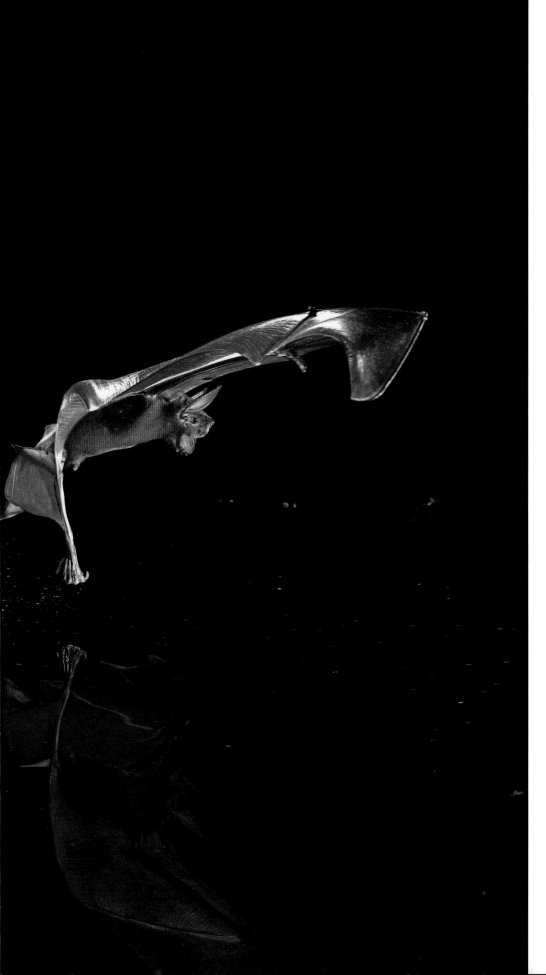

Killer whale pool tricks

Looking out at the South Atlantic, you get only a hint of the power of this massive body of water. The huge grey waves that crash onto the shore seem loaded with pent-up energy. Water is the dominant element here. But at certain times of year, the few islands that rise above the storm-blown seas become magnets for the wild inhabitants of the South Atlantic.

One such landmass is Sea Lion Island, part of the Falkland Islands. It's long and thin, and to the north lies a saddle of windswept sand and ridges of old peat beds. Here there are colonies of gentoo penguins that nest above ground on the large mounds spread across this area. There is a beautiful symmetry to these colonies. When the wind is really howling, the penguins are like weather vanes – each lies facing away from the wind, contributing to the appearance of a flock flying in formation.

Leaving every morning and evening is a steady stream of penguins heading down to the beach and out into the surf to feed in the open ocean while their partners guard the nests. Each evening there is the spectacle of 'flights' of penguins porpoising back through the waves, travelling fast as they race to catch a wave on which to surf back to shore.

Below Fat elephant seal pups lying around their nursery pool on Sea Lion Island in the Falklands. Here they practise swimming, in preparation for life in the ocean. But at high tide, a channel links the pool to the sea – a way in for a predator.

It soon becomes apparent why they return in such haste. Cutting through the water behind them come huge black fins – killer whales, the largest and most predatory members of the dolphin family. Around Sea Lion Island are one or maybe two pods of killer whales. In November and December, they are here for bigger prey than the tiny penguins – southern elephant seal pups, strewn across the beaches of the island.

Back in September and October there was a riot of activity here as the beaches filled with male elephant seals fighting for breeding rights and females coming ashore to give birth and mate again. But now all that is left of the party are the pups. Born, suckled and abandoned by their mothers in just three weeks, the pups will stay here until December or January, when their increasing maturity and hunger will send them out into the open sea.

For now they lie along the shoreline, alternately sleeping, play-fighting and swimming in the shallow sea pools. Mornings are when they are most active, venturing into the sea pools that mark the southern end of the main beach. One particular pool is a real favourite. Lined with a swathe of kelp, it's the perfect nursery and training ground for the seals.

Above *Pups splashing around at high tide, unaware of the silent entrance of a female killer whale and her calf along the channel and into the nursery pool.*

Knowing the layout of this pool is crucial to understand what follows. It's roughly the size of a municipal swimming pool. Carved out of the bedrock, it's quite shallow on the landward side but shelves down towards the sea. On the sea edge is a ridge of rock that runs right across the front of the pool, enclosing it.

But cutting right through the middle of the sea-defence ridge is a channel connecting the pool to the sea. It's 35-45 metres long (115-130 feet), and at low tide, it's neither very deep nor very wide. Occasionally pups will check out the mouth of the channel but almost never venture into the channel itself, preferring the shallower waters of the pool. For such aquatic mammals, they seem nervous of deeper water. And they do well to be nervous. Not only do the killer whales know all about this pool, they also know how to get into it.

What it was that first alerted the killer whales to this small pool we will never know. Given that the noise from these seal colonies is both constant and carries a long way, it may be that sound played its part in attracting the killer whales to the mouth of the narrow channel. It might even be that the killer whales learnt about this particular pool from their own mothers – that they have hunted here for generations.

Arriving at the outer reaches of the rock ledges before dawn, the pod is silent as it cruises ominously along the shore. Normally it's possible to see killer whales from a long way away by the spume that erupts from their blowholes as they surface to breathe. But now each exhalation seems muted, and they surface less regularly than normal. While most of the pod circles just offshore, one of the females, followed by her calf,

swims towards the mouth of the channel. There is a terrifying focus in this approach, and the killer whale is soon inside the channel. She swims slowly, as if sensing the push and pull of the waves, judging whether conditions are safe for her and her calf. This is an enormously dangerous strategy – if she becomes beached by a big wave, there would be no hope for her.

She edges forward quietly through the channel and into the pool itself. Here she pauses, hanging immobile in the water. Maybe she is using her sonar to judge if there are pups in the water. But what is certain is that the underwater visibility in the pool is nearly zero, as the tide constantly churns up the mass of vegetable matter. After hanging still for as much as a minute, she turns and makes her way back out into the open water, followed by her calf. It seems that she cannot stay too

long in the pool – probably because the waves push her in bit by bit. Time and again this process is repeated: coming into the pool, waiting and then turning and leaving. Each time the calf follows its mother – perhaps, in the process, inheriting the knowledge that will allow it to continue this strategy in adulthood.

Finally she gets lucky. As she comes into the pool, one of the seal pups that have been playing in the shallows decides to swim across the pool. As it swims along with its head out of the water, it suddenly becomes aware of the strange black shape. Wary but interested, it swims closer and then stops.

The killer whale is fully aware of the pup, and sensing its hesitation, she slowly rises in the water and exhales the gentlest of breaths. This simple, familiar

Above *A curious pup swimming up to investigate the human in its pool. It will do the same when a killer whale approaches, with fatal consequences.*

Above An elephant seal pup catch. Once the pup is in open water, it is released. Then one of the killer whale pod slams into it, hurling the pup out of the water and inflicting deadly injuries.

Opposite Patrolling killer whales preparing for another silent entry into the nursery pool. The opportunity for such rich pickings comes only once a year. The hunters must not only remember the place and time but also learn and perfect the technique for making a kill.

sound seems to reassure the pup, and it starts forwards again, its path set to cross some three or four metres in front of the killer whale. Just as it passes an invisible line, the killer whale rushes forwards. The pup has no time to react. It is plucked from the surface like a fly taken by a trout, and the pool erupts in spray as the killer whale thrashes her tail, driving herself around, trying to reach the channel. Propelling herself and 120 kilos (265 pounds) of seal pup forward is no easy matter. By now the pup is flailing about, trying to escape, as the killer whale drives herself forward in a series of lunges. Finally she reaches the channel and soon the open sea.

As she enters the deep open water, the rest of the pod rushes to meet her, and soon there is a mêlée of fins and backs, flashes of black and white shapes in the water, rolling and porpoising over and over.

But miraculously, it seems that the pup has escaped, as with its head bobbing out of the water, it starts to swim back towards the shore.

But this is no accident – the killer whales have let it go deliberately. This seal pup is a big animal. It has sharp claws and a mouth full of teeth. The scars around the faces of the killer whales are testament to the damage their prey can do. So to minimize the risk, the killer whale releases the pup, allowing it to swim away. One killer whale then swims at incredible speed straight at the pup's flank, slamming into it with such huge force that it is hurled out of the water. The damage inflicted is enough to finish the pup off. Once the pup is eaten, the female returns to the pool, and the whole process begins again. In a four-day period, eight pups might be taken – a massive amount of protein for the pod.

And the female is not just reliant on a pup swimming past. If a pup is on the ledge at the edge of the channel, she will generate a wave by rocking backwards and forwards in the water, trying to wash it off the rocks. If this fails, she'll swim parallel to the rock, and as she comes up to where the pup is lying, she'll roll onto her side attempting to sweep it off with her dorsal fin.

It is extraordinary that the killer whales have learnt how to exploit this resource, given that the opportunities are so incredibly limited. To hunt like this, conditions must be perfect – a calm sea, a high tide in the morning and, crucially, the presence of seal pups. There might only be five mornings a year when the killer whales can actually hunt like this. Yet they have learnt to revisit this location year after year and to exploit the fleeting opportunity. The success of mammals as hunters is very much based on this ability not only to take advantage of brief hunting opportunities but also to revisit those locations and to remember the how, where and when that made them successful in the first place.

The art of seal tipping

Killer whales are truly global: no other mammal ranges so widely through the seas. They are also one of the most intelligent and sophisticated of hunters – as eclectic in their hunting methods as they are in their choice of hunting grounds.

In winter, the Antarctic is no place for animals. On the Antarctic Peninsula, glaring white glaciers sweep down through towering peaks of ice-clad mountains to meet the frozen sea, and even in September, the sea ice keeps a grip on the bays and inlets. But as spring progresses, the sun's influence is felt. The ice begins to break up into ever-smaller pieces, and once more, icebergs are released to drift. Gradually the peninsula opens up, and the penguins, whales and seals penetrate the calm backwaters of its bays to find food and shelter. In this arena of exquisite beauty and cold, killer whales are the top predators. There are three types here: one that feeds mainly on whales, particularly minke whales; another that specializes in fish; and a third that prefers seals. This last group has a grey rather than black appearance, with white patches tinged yellow by a patina of diatoms (a type of phytoplankton). These killer whales also have an unusually large eye patch running parallel to the body.

There are seals in abundance around the Antarctic Peninsula, from fur seals and Weddell seals to leopard seals and, most numerous of all, crabeater seals. All rest on the ice, and as the ice breaks up, they can be seen alone or in large groups snoozing on the drifting floes. Their snoozing is, of course, intermittent, because they know that they are not entirely safe. And as spring turns to summer and the floes become smaller, thinner and less robust, the danger increases.

Explosive breaths and large dark fins breaking the surface warn of the passage of a pod of killer whales through the ice-free channels. Every so often, one killer whale will spy-hop, looking for resting seals. When a seal on a small-enough ice floe has been spotted, the ice-tipping will begin.

Above *A resting crabeater seal – favourite prey of certain communities of killer whales in the Antarctic.*

Right *Cruising seal-killers. Their skin is more grey-black than that of other killer whales and their white patches are stained yellow by diatoms (microscopic algae) living in the sea where they hunt.*

Next page *Crabeaters on sea ice. Though confined to the Antarctic, they are the world's most numerous seals, and in spring, at least 14 million haul out to give birth – providing fodder for leopard seals as well as killer whales.*

Above *Two killer whales spy-hopping to determine the ease of getting a meal and how to wash it into the water.*

If the ice floe is too big to tip up – say, 20m in diameter – the whales engage in a series of manoeuvres to reduce its size. Two or more will swim fast at the floe from some distance and dive just in front of the ice edge, which creates a wave that washes over the ice floe. It's a technique that may or may not wash the seal off into the water, but what it is intended to do is break the ice into smaller pieces,

making the seal more exposed and easier to dislodge. The whales push the smaller fragments of ice away from the immediate area by blowing bubbles and by diving to cause more turbulence. If broken ice still surrounds the beleaguered seal, the whales may even use their rostrums to push the floe to ice-free water.

By now, the seal will be highly stressed – probably hyperventilating and working its jaws. But there is nothing it can do. To enter the water would be suicidal, and so it clings to the ice.

The hunters' final manoeuvre varies from pod to pod and according to the conditions of the ice floe or the sea, but when the ice floe is some 5m (53 feet) in diameter and small enough to tip, some of the whales will rush at the floe, swimming on their sides. As they reach the edge of the floe, they will dive under it and, at the far side, turn abruptly and wait under the surface. (It's thought that the whales swim on their sides so they can swim close to the floe and pass beneath it without damaging their dorsal fins.)

The seal has little chance. First the ice floe tips towards the trough in front of the bow wave created by the whales as they approach the floe. Then it rises and tips the other way as the wave lifts it. The breaking crest of the wave washes the seal off the floe directly onto the waiting whales.

The seal rarely meets a quick, clean death. In most cases, a killer whale will catch the seal, swim for a while with it in its mouth and then release it a number of times before one of the whales finally kills it. The whales may even place the seal back on an ice floe before washing it off again. This is probably partly training to perfect technique and partly a way of teaching the younger whales in the pod how to hunt.

Sometimes a seal does escape, gets onto the floe again and is left alone, though it may not survive internal injuries sustained during the hunt. Seals can also use blocks of ice for protection. In one recorded incident, a pod of whales was unable to catch a seal that swam round and round a block of ice. After 40 minutes of continuous and violent porpoising very close to but never actually touching the ice, the killer whales gave up, leaving the exhausted seal cowering behind the ice.

Hunting that involves such a coordinated approach is rare in animals. Members of the dolphin family, of which killer whales are the largest, demonstrate the most sophisticated techniques of all, and it would seem that they can perpetuate their strategies – which vary from region to region – passing the knowledge from one generation to the next. This so-called 'culturally transmitted hunting behaviour' exhibited by killer whales is further evidence of the advanced intelligence of these animals.

Left *Moving in for the kill. After weighing up the size of the ice floe, the killer whales opt for the synchronized technique. Swimming in formation on their sides at and then under the ice, they create a wave that tips the ice and then washes off the seal.*

African wolves with altitude

Above Ethiopian wolves setting off on the early morning patrol. Though they will hunt alone, all members of the pack work together to beat the bounds and warn off other packs with yelps, barks and howls.

Opposite A nanny left behind at the den to look after the pups. Though only the alpha female breeds, all the pack members help look after the pups, returning at intervals to regurgitate the rodents they've eaten.

We tend to think of Africa as being covered by rolling grasslands, thick jungles and vast deserts. But there are parts of this fabulous continent that are very different. Ethiopia's wildlife is dominated by its highlands, and living way up on this spectacular high-altitude dome are some truly unexpected predators – none more surprising than the Ethiopian wolf.

Even before Western science had discovered the Ethiopian wolf (and then misnamed it a couple of times), the species was rare. Today, with a total population of fewer than 500 individuals, it is quite probably the world's rarest canid. The various populations of wolves are isolated from each other – each trapped up on a separate mountaintop, increasingly at risk from human encroachment and the diseases carried by domestic dogs.

Sharing a common ancestor with the grey wolf, the Ethiopian wolf came to Africa some 100,000 years ago, during the course of a vast ice age. When the ice retreated, the wolves remained up on the very highest mountains. They have retained the pack structure of their ancestors. The alpha male and female are the only ones that breed, while other members of the pack help raise the pups. There is a twist here, though – not all the pups will necessarily have been fathered by the alpha male, because the alpha females have been known to mate with males outside of the pack.

Living in a pack is beneficial in a number of ways. It not only helps in the raising of young but also makes it easier to protect and patrol a large territory, which may be up to 13 square kilometres (5 square miles). And this is normally the first job of the day. Having spent a very cold night curled up between helichrysum bushes, with their tails around their noses for warmth, the wolves wake up to the frost of an Afroalpine morning. As they rise and stretch, they greet each other, and soon there is a joyous writhing mass of wolves, reaffirming the bonds that keep the pack a cohesive unit. Soon they are ready to head out on patrol.

Travelling in a loose formation, the wolves roam across their territory checking that there are no interlopers on their patch. Should they spot a neighbouring pack, the dispute is normally resolved with yelps, barks and howls rather than by physical attack. Having beaten the bounds, individuals head off to hunt.

While grey wolves will hunt alone only when necessary, Ethiopian wolves are lone hunters. There simply isn't sufficient prey here large enough to make pack-hunting worthwhile. But what the prey lacks in size it makes up for in numbers. These highland regions are home to incredible numbers of rodents – grass rats, mice and, most important for the wolves, mole rats. In some areas there are up to 2900kg (6400 pounds) of rodents per square kilometre – an extremely rich food source but one that is very hard to catch.

But the wolves are master rat-catchers. Once a rat has been spotted out of its hole, the wolf edges forward, often with its belly flat to the ground, trying to close

the gap between it and the rat without being spotted by the target or any of the other countless alarm-givers about. Every time the rat looks away, the wolf closes in by bunny-hopping forward in short lunges or by making short rushes, until it gets close enough to pounce. Lunging forward, it attempts to grab the rat. But these rodents are incredibly quick, and often a hunt ends with a scramble, as the wolf tries to grab a rat that is now racing around trying to find a hole to hide in. With luck and a fair amount of dexterity, the wolf snatches the rat and tries to dispatch it with a bite, being wary not to allow the rat to turn and deliver a bite of its own to the wolf's sensitive face. If the rat manages to dive down a burrow, then the wolf will attempt to dig it out.

When there are pups back at the den, the wolves return in a steady stream during the day to regurgitate the rats they have caught. These are snapped up by the newly weaned pups as well as the 'nanny' left behind to guard them. In this way the wolves are able to maintain a foothold in this harsh environment.

But the wolves now face far greater challenges than simply finding food and raising their young. As in so much of the world, land is increasingly under pressure from an ever-expanding human population. With people come their dogs, and with the dogs come disease. The wolves are closely enough related to domestic dogs to be susceptible to the same diseases, and in such a social species as the Ethiopian wolf, disease spreads shockingly fast. Up to 80 per cent of one population was wiped out in 2003 by rabies – spread by domestic dogs.

So while the wolves have found a home high on Africa's mountaintops, there remain massive challenges threatening their survival. None of the wolves' incredible hunting or social abilities will help them overcome these challenges. That is for us to do – to ensure that, for generations to come, these fabulous hunters continue to stalk the mountains.

Above *Rat-catching. The initial rush and pounce has failed to catch a fast-moving mole rat – the favoured prey – and now the wolf is attempting to extract it from its hole.*

Left *A family group warming up in the morning sun before setting off on patrol.*

Mud, mullet and the mammal marines

Above *A bottlenose dolphin plucking leaping mullet, having herded the shoal against the shore. Individual populations in Florida waters have devised different cooperative ways to fish.*

The Florida Keys are a curious mix of mangroves and small islands. Between them are vast areas of mudflats covered in a shallow skim of water. Flying over these areas reveals intriguing marks in the mud – open loops that, like crop circles, slowly fade as the wind and the tide washes them away. These mud rings are the traces of a most extraordinary hunting strategy – one that involves special senses, teamwork and the ability to manipulate the environment.

Dolphins are well known for their intelligence, and the Florida bottlenose dolphins – or at least certain pods of them – have developed a unique way of catching their prey. This population has a number of problems to solve. First, the very richest feeding areas are in the shallows, and so the dolphins are forced to swim along channels to reach them, often swimming on their sides as they edge along to the slightly deeper areas beyond.

Once they have accessed the flats, their next problem faces them – finding their prey. On the flats live large shoals of mullet. Like all dolphins, the bottlenose use sonar to locate prey – firing off a series of clicks and then listening for the resulting echoes. The moment a shoal is spotted, one of the dolphins will race towards it at full speed, surging through the water.

As it comes up on the shoal, this lead dolphin swims in a perfect circle around the mullet – ending up pretty much where it started. But as it goes around the shoal, it beats its tail powerfully downwards. This violent action drives the mud up from the seabed, immediately creating a wall of muddied water that now surrounds the mullet.

As the dolphin reaches the top of its loop, a colleague races in to join it, and they line up, absolutely side by side, just outside the mud ring. At this point the mud ring starts to lose its perfect shape and collapses in on itself. The mullet trapped inside panic at what seems to be danger closing in on them from all sides, and to escape it, they leap from the water, trying to jump over and out of the closing circle.

But this is exactly the response the dolphins are after. By now they have their heads up out of the water, and as the coveys of mullet rocket from the water, the dolphins snatch them from the air.

Like cricketers practising their slip-catching, the dolphins leap athletically to pluck the fish from the air – sometimes arching right back to catch a particularly high-jumping one, sometimes going out to full stretch one way and then another to catch fish just before they regain the safety of the water. In seconds, the mud ring has totally collapsed, and the mullet that haven't been caught have raced away.

The dolphins move off, scouring the flats ahead for more shoals. Very soon, a mud ring starts to develop as, once more, these highly intelligent, adaptable, social mammals put on an extraordinary display of teamwork.

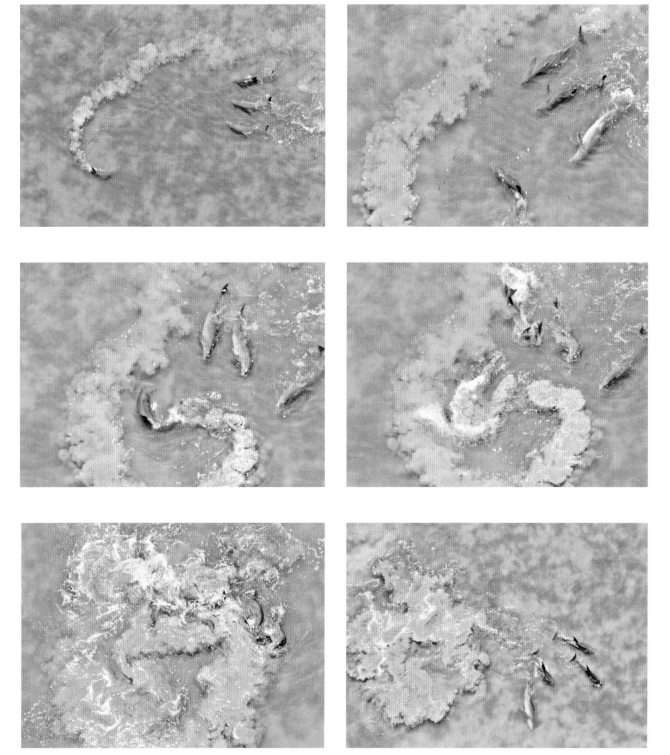

Left *A pod of Florida Keys bottlenose dolphins practising the mud-ringing technique, perfected for mullet fishing in the shallows. One dolphin encircles the fish, beating its tail to create a wall of muddy water. As the mud closes in and the mullet panic, they start to leap – into the mouths of the athletic, high-jumping dolphins.*

chapter **9**

Intellectual primates

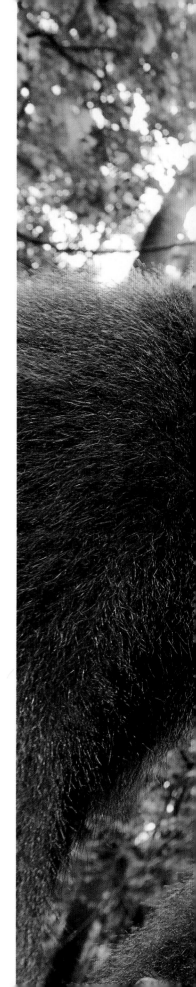

WE MAY NOT APPEAR TO HAVE much in common with the pygmy mouse lemurs of Madagascar or the acrobatic lar gibbons of Thailand, but like them, we are primates. We shared a common ancestor back in the time of the dinosaurs, and today, we are one of about 635 species and subspecies of primates belonging to this extremely successful order.

There is no single defining feature that sets primates apart from other animals – more a range of distinctive traits heavily influenced by a tree-living lifestyle, which most possess. Forward-facing eyes with stereoscopic vision give us a sharp depth perception and a three-dimensional view of the world – crucial for life in the treetops, where our ancestors evolved. Having five digits on both feet and hands, with an opposable thumb that can reach all the other fingers, gives us great dexterity for getting about, handling objects and using tools. Though humans exchanged opposability in our big toes for walking on two feet, chimpanzees still retain theirs and can manipulate objects with their feet. All primates also have nails, rather than claws, which help to protect fingers and toes on one side while allowing enhanced touch sensitivity on the other. Perhaps most significant, though, is that primates have evolved larger brains than other mammals of similar size, particularly the neocortex area, which makes up 50-80 per cent of brain volume and is responsible for abilities such as consciousness and reasoning. This could have happened for a mix of reasons, ecological as well as social. But most primate brain development happens during the extended social period between weaning and adulthood, when there is so much to learn from experience.

Primates are divided into two groups (suborders), the Strepsirrhines, which include lorises, lemurs, bushbabies, indris and aye-ayes, and the Haplorrhines, comprising

Right The world's smallest primate, the pygmy mouse lemur. It's nocturnal and tree-living but shares many characteristics with humans, including a large brain, forward-facing eyes with stereoscopic vision and opposable thumbs.

Opposite Adolescent crested black macaques displaying a very primate characteristic – curiosity. All primates have an extended childhood, giving them time to learn from experience.

Previous page Geladas grazing and socializing. Most primates live in tropical or subtropical forests, but these are confined to the Ethiopian Highlands.

tarsiers, monkeys and apes. Today's primates have a great range of fascinatingly different social structures, moulded by many factors, including the environment where they live, the food they depend on, threats from competitors and danger from predators. Some, such as orang-utans, lead a solitary existence, with mothers caring single-handedly for each infant for up to eight or nine years, whereas others, such as lar gibbons, form stable male-female partnerships, with both parents looking after the young. Western lowland gorillas live in extended family units governed by a silverback male leader, while Japanese macaques go a stage further, forming multi-male, multi-female troops full of intricate social relations and hierarchies among females. One of the most multifaceted of all social systems, humans aside, is that of hamadryas baboons. They live in small, single-male-led units composed of females and their young and one or more 'follower' males but which associate with other one-male units to form foraging and sleeping troops of up to several hundred baboons.

Non-human primates are found as far north as Honshu in Japan and as far south as the Cape in South Africa, but most live in tropical or subtropical forests and depend on a year-round supply of food. Primates eat a whole range of items including insects, frogs, crabs and other mammals, but for most, a staple diet includes leaves, roots, seeds and fruits (gathered with the aid of colour vision). And while they feed, primates play a crucial role in maintaining the health and diversity of forests by helping with seed dispersal, soil fertilization, pruning and control of insect pests.

All primates spend an extended childhood dependent on their mothers for warmth, security, transport and education. They must learn when and how to find food, who to trust and not trust, how to best communicate with others through scent, sounds, touch and sight, and what and who to beware of. This level of parental care and investment can sometimes take half a lifetime, as is the case with humans, and is what truly sets primates apart from all others.

Like us, many primates also have distinctive local cultures handed down from mothers to infants. Some of the most fascinating concern tool-use. In the Cerrado savannah of Brazil, bearded capuchins regularly use heavy rocks to crack open palm nuts. In Sumatra, orang-utans select twigs to probe for and extract honey and insects. And in Bossou in Guinea, chimpanzees use oil-palm leaf stalks as pestles to pound the juicy pith of the plant.

While our knowledge of and respect for other primates is growing, so their numbers are falling. For no matter how advanced their societies are or adaptable their lifestyles, they cannot compete with us. Deforestation for timber, agriculture and settlement, along with hunting and disease, means that nearly 50 per cent of species, including our closest relatives, are listed as threatened and could go extinct during our lifetimes.

Above A ring-tailed lemur feeding on aloe blossoms. Colour vision helped primates to take a bird-like niche in the forests, feeding by day on flowers, fruits and seeds. It also provided an opportunity to use colour for social and sexual signalling.

Left Howler monkeys calling at dawn from the canopy, their prehensile tails functioning as extra arms. Group roaring – among the loudest calls made by any primate – carries more than a kilometre, communicating their whereabouts to other groups and individuals, warning them off.

Moonlight hunters

Above *A leap and snatch resulting in a catch for the spectral tarsier.*

Opposite *A foraging tarsier. Huge, movable ears help the night-hunting tarsier pinpoint the sound of prey, and enormous eyes gather whatever light is available for closer focus. Unlike most other nocturnal animals, it doesn't have a light-reflecting tapetum and has retained colour vision, pointing to a closer relationship to monkeys and apes than to aye-ayes and lemurs.*

With eyeballs each the size of its brain – so big they are hard to move within their sockets – a neck so flexible it can rotate its head almost 360 degrees and feet with such elongated tarsus bones (hence its name) that the ankle hinges twice, a tarsier is an extraordinary primate. It is also the subject of much scientific debate.

Tarsiers share many traits with the more primitive group of primates known as Strepsirrhines, including lorises, lemurs, bushbabies, and aye-ayes, most of which are nocturnal. But tarsiers lack some of their fundimental features, in particular, the tapetum lucidum (the light-reflective layer at the back of the eye) and the wet nose (an aid to smell) – both adaptations for night life. They also have colour vision similar to New World monkeys. These differences have resulted in tarsiers being classified along with monkeys and apes, including humans.

But though they lack some of the more usual nocturnal attributes of Strepsirrhines, tarsiers have a nocturnal lifestyle. They live in forests on islands of the Southeast Asian nations of Brunei, Indonesia, Malaysia and the Philippines, and are among the smallest primates in the world, with a head and body just 10-15cm (4-6 inches) long, though their hind limbs are almost twice this length. What really sets them apart from other primates is the fact they are rapacious carnivores, eating no plant matter whatsoever and relying entirely on acute hunting skills.

At Tangkoko Nature Reserve on the Indonesian island of Sulawesi, spectral tarsiers live in groups of two to ten individuals – most usually a male and female and their offspring. They prefer to spend the day sleeping high up among the aerial roots of strangler figs and emerge just as darkness falls. Like all other tarsiers, they are vertical clingers and leapers, able to leap several metres between saplings or trees with amazing proficiency and accuracy, powered by their long, muscular hindlegs. Their feet are the first to hit the landing site, followed by their hands with their especially long fingers. They can also climb, hop and walk on all fours. Managing precise arboreal feats at night depends on the sharpest vision possible, and lacking a tapetum, tarsiers rely instead on their enormous eyes and dilated pupils to gather every photon they can from the moon and stars. If the moon happens to be full, they become especially active.

Spectral tarsiers spend more than half their waking hours foraging, and for their size they travel great distances in search of food within a home range, in Tangkoko, of up to 4.1ha (10 acres). They defend their territories with vocal duets, and scent-mark tree branches as they go. An infant travels with its mother. When it's very young, it's carried in her jaws and 'parked' while she forages; after that, it clings to her fur until it's about 45 days old and can feed on its own. Travelling a metre or two off the ground offers some degree of safety, but spectral tarsiers also sometimes

forage on the forest floor, especially in the dry season, when food is harder to come by. Being small, it has many potential predators, including monitor lizards, snakes and Malaysian civets and so is constantly on the alert. If danger is spotted, a tarsier makes an alarm call, and group members may join in to mob the predator.

A tarsier doesn't just look but also listens for its prey – mainly insects such as beetles, cicadas, moths, caterpillars, crickets, katydids, grasshoppers and cockroaches, as well as spiders, termites and ants – aided by its swivelling head and ever-moving sensitive ears. Most insects are snatched from leaves and branches or pounced on, but a tarsier's vision is so good that it can also grab prey in mid-air.

That tarsiers have evolved such an original lifestyle is intriguing (the only other nocturnal haplorrhine primates are the South and Central American owl monkeys). It is probably that, towards the end of the great dinosaur era, the ancestors of tarsiers became diurnal (day-living), as did most monkeys and apes, and lost the tapetum. The fact that tarsiers have colour vision seems to support that theory. Specializing in eating insects and other prey animals active under cover of darkness may then have encouraged a return to the night and a plentiful supply of food. Lacking the tapetum, they opted for an owl-like strategy, developing their huge, forward-facing eyes, rotating head and acute hearing. With seven, possibly more, species of tarsiers in existence, it has proved to be a hugely successful strategy.

Above *Travelling in search of food. Tarsiers can easily leap three metres between trees, using their long, muscular legs, landing feet-first and grabbing with their especially long fingers, which have rounded pads at their tips.*

Left *A group of tarsiers emerging from their communal daytime sleeping place in a fig tree.*

Family life and the fruit factor

Opposite Mother and child in the dense rainforest of the Congo. A baby will stay close to its mother until it is at least three years old.

Gorillas – among the most threatened of primates – were officially 'discovered' and described for science only in 1847, when a missionary brought back a gorilla skull. Found in West and Central Africa, western lowland gorillas inhabit dark and dense rainforest. Coming upon them in such circumstances is as frightening for humans as it is for gorillas, and either party is likely to flee or become defensively aggressive. So it's not surprising that lowland gorillas

(397 pounds). He also has large canines. The silver saddle across his back is a badge of maturity, developed at about 14 years old. Until then, males have uniformly black fur. Females are half the weight and size of silverbacks and may have distinguishing reddish brown hair on the tops of their heads.

Western gorillas live in stable family groups typically consisting of one silverback, three or four females and

Above Youngsters play-fighting. Play is a hugely important part of primate development, allowing experimentation and the development of social skills.

Above, right A male beating his chest, either to make contact with his females or as a proclamation of dominance. Despite their size, gorillas are remarkably unaggressive, using social skills to avoid violent confrontations.

were little observed in the wild until the early 1990s, when zoologists discovered open areas of marsh – *bais* – where gorillas come to feed on water-plants rich in sodium. The plants have a high water content and can be gathered easily, which means the gorillas reach their fill after only an hour or two and then head back into the forest. So, even today, we know comparatively little about their family life.

The role of an adult male gorilla, the silverback, is family protector. He stands no taller than a human, but he may weigh three times as much, averaging 180kg

four or five youngsters. By day, the group may spread out. Quiet grunt-like sounds maintain contact. At night, they come together for protection, sleeping on the ground (in trees only if it's wet). Fruit forms a large part of a gorilla's diet and, being seasonal and patchy, restricts the size of family groups. A small group can feed around a fruiting tree without much competition and fighting, whereas a large group could not. But if the area a group occupies contains many fruiting trees, a larger group becomes possible – one with nine females has been observed. But it's not just the availability of fruit that determines group size.

Females will stay with a silverback they deem strong enough to protect them and their infants from leopards and from other silverbacks who might kill their babies. Today, though, the greatest threat comes from trophy and bushmeat hunters with guns and the Ebola virus.

On reaching maturity, a female will find another group to join or, occasionally, a lone male to team up with. Young males almost always strike out on their own. When they are large enough and strong enough, they will attract females. The sound of a silverback's chest-beating resounds through the forest. It can be used to keep in contact with the females, but it may also signal trouble. Gorillas on neighbouring patches usually tolerate each other, but should a lone silverback enter a family territory, a fight breaks out. It is ritualized to avoid serious injury, starting with hooting, rising to a crescendo and ending with chest-beating. As a grand finale, the ground is thumped or a branch torn from a tree.

The current belief, though, is that one silverback rarely takes over a group from another. When a silverback dies, his females may be adopted by a nearby young male, but unless he impresses them, they will split up and find other groups to join. The ease with which females migrate and the lack of mutual grooming observed in the *bais* indicates that female bonds are weak, though there is a hierarchy among them.

The bond between infant and mother is, by comparison, very strong and remains so for many years. For at least three years, a youngster continues to suckle and stay close to its mother, riding on her back as she moves around, learning what to eat, how to avoid trouble and how to behave around other gorillas. This long learning stage limits the number of offspring a female can have in her lifetime. But the investment is necessary if a young gorilla is to learn all it needs to for survival. Youngsters will play a great deal together but don't appear to form long-term relationships, probably because, when they reach maturity, they leave the group and set out on their own.

Left *A mother feeding in an open marshy area on water plants rich in sodium. Her young infant watches carefully what she eats and experiments with new tastes. These clearings are among the few places where lowland gorillas can be observed outside the thick forest.*

Opposite *A youngster developing its tree-climbing abilities. Like other social apes, including humans, it has an extended childhood in which to learn social and other skills.*

Next page *A dominant silverback male and father of the infants in his group, his red head marking him out as an adult.*

The educated ape

Every morning, Gunung Leuser National Park in northern Sumatra, Indonesia, comes alive with a cacophony. The booming calls of rhinoceros hornbills echo across the steaming valleys; a pair of white-handed gibbons begin a whooping, wailing duet; and the drone of barbet song and cicada strumming builds in intensity. Then, from the canopy, come the 'long calls' of the world's largest tree-dwelling animal, the orang-utan.

Orang-utan means 'person of the forest' in Malay, and these Asian great apes are found only on the islands of Sumatra and Borneo. Though they can weigh as much as 90kg (198 pounds), they are canopy experts and supremely mobile, using measured, 'quadrumanual' climbing. Big toes allow the feet to grasp as firmly as hands, and hip joints ensure great pivotal mobility. This arboreal prowess, together with learning skills, enable orang-utans to select the best canopy routes, whether these involve acrobatically swaying back and forth on tall saplings with outstretched arms to grab neighbouring branches, swinging on lianas (vines) or scaling trunks to find the best branch-bridges. If a crossing is too wide for a youngster, its mother will use her body as a bridge.

The sheer weight of an orang-utan means that, as it moves, there's a fairly constant barrage of falling branches. And though at least two limbs keep a grip at any one time, occasionally an orang-utan will fall and break bones. It begs the question, why don't orang-utans spend more time on the ground, like gorillas and chimpanzees, selecting fruiting trees to climb, rather than undertaking such risky effort in the canopy? In Sumatra, the obvious answer is tigers and clouded leopards, but health is another consideration. By avoiding the forest floor, there is less chance of picking up parasites such as protozoans and intestinal worms through contact with faeces and contaminated soil.

Even more impressive than the arboreal skills of orang-utans is the extent of their motherly care. Females become sexually mature around the same age as human girls do, and pregnancy lasts eight and a half

Above *Learning what's good to eat. A young orang-utan may spend up to eight or nine years being educated by its mother about life in the forest – the longest child-rearing period of any non-human mammal.*

Right *A female from the Ketambe region, plucking seedpods from a tree trunk high up in the canopy – one long arm holding onto a safety branch. Both activities reveal just how important grasping hands are.*

months. The female may then spend the next eight or nine years as a single mother raising her infant and teaching it the ways of the forest. It's the longest birth interval of any land mammal (humans included), and the childhood is the longest of any non-human animal.

There is a considerable amount to learn in the world of a Sumatran rainforest. In the Ketambe region of Gunung Leuser National Park, 35 years of research have revealed that mother orang-utans teach their infants to source fruits from almost 200 tree and liana species, with figs a strong favourite, in a home range that may stretch to 4.5 square kilometres (1.7 square miles). For a rich and varied diet, they have to be taught to find particular leaves, flowers, pith, fungi, honey and termites, and even how to catch small mammals they may chance upon, such as slow lorises. An infant's education also includes the construction of day and night nests, sun shades and umbrellas, and gloves out of leaves when feeding in a spiny tree.

Above *A female looking for honeycombs and termite nests. Orang-utans in another area have been seen using tools to extract honey or insects.*

Left *A male, nearly full size. When fully mature, he will have even broader cheek pads and a larger throat pouch. Males don't reach maturity until about 18, and even then, their reproductive success depends on competition with other, more mature, males.*

Above *A young orang-utan tasting the fruit her mother is eating. She must learn how to recognize nearly 200 different kinds of edible fruits and also what else is edible and how to find, extract or catch it.*

Opposite *A mother and her infant. Child-bearing, as for humans, is extremely demanding, with a pregnancy of eight and a half months and eight or nine years of child-rearing – the longest of any non-human primate.*

There are cultural differences in behaviour. In the Suaq lowland-swamp-forest region of Gunung Leuser, where females occupy larger home ranges – up to 8.5 square kilometres (3.3 square miles) – they use branch scoops to gather water from tree holes, twigs to probe in crevices for insects and stingless-bee honey, and stripped sticks to dislodge the seeds from neesia fruits, which are covered in irritant hairs. Swamp-forest orang-utans even have different vocalizations from the Ketambe ones, such as blowing through pursed lips to make a phhhhp – a sort of raspberry – on completion of nest-building.

Though all orang-utans lead a solitary existence, every once in while they will unite, either as male-female liaisons or to feed. Some of the biggest gatherings of Sumatran orang-utans can be seen when fig trees are fruiting, and during the mast season, when many trees fruit simultaneously. Orang-utan fruit-tree gatherings can be extremely sociable, with close relatives feeding together and endless play among the youngsters.

For young orang-utans branching out on their own, this extensive education, handed down from generation to generation, is fundamental to their success. During their lifetimes, they will be key contributors to the biodiversity and renewal of the rainforest, especially through their seed dispersal. And because of their dependence on this habitat, they are also crucial barometers of its health. But with continuing pressure from illegal logging, forest fires, the rapid spread of oil-palm plantations and illegal hunting, fewer than 6600 of these highly intelligent great apes remain on the island of Sumatra.

Ways to keep warm

Macaques are the most wide-ranging non-human primates, with more than 20 species, found from North Africa across to the Himalayas, southern India and Southeast Asia, and they live in habitats as diverse as tropical mangrove swamps and mountain cedar forests. The hardiest and most northerly is the Japanese macaque, which has developed to survive winter temperatures that can drop below -20°C (-4°F).

The Japanese macaques that live in the forested mountainous areas of Honshu, the largest island of Japan, are nicknamed snow monkeys. They are stocky with thick fur, and live in troops of 20-100 individuals. Females and their young heavily outnumber males by three to one or more, and each troop contains several matrilineal groups with rigid, inherited hierarchies, infants being bestowed with the ranks of their mothers. In parts of their range, winters can be particularly severe, with heavy snowfall as well as freezing temperatures. In such conditions, being of high rank really matters when it comes to keeping warm and finding enough to eat.

For most of the year, they eat mainly fruit, but in winter they have to be flexible, foraging for relatively poor-quality food such as tree bark, winter buds, plant roots or bamboo grass, supplementing this where possible with protein-rich insects and larvae, nuts and fungi. Higher-ranking animals monopolize the best food patches and so have more chance of getting enough calories and protein. Fat reserves laid down in times of plenty are vital to top up such a meagre diet. A layer of fat, along with a thick winter coat, also provides insulation, and the monkeys will huddle together for extra warmth, curling up their toes to try to prevent frostbite.

In the mountains of northern Nagano, in a place called Jigokudani (meaning Hell's Valley), the Japanese macaques have found another way to keep warm. Japan sits on the Pacific 'Ring of Fire', its mountainous backbone peppered with highly active volcanoes.

Above *A high-status Japanese macaque mother suckling her infant while keeping warm in a hot-spring pool. Lower status macaques may not be allowed access to a pool.*

Left *A high-ranking infant playing in the 'hot tub'. It has inherited its status from its mother – giving it a major advantage in freezing conditions.*

Above *Warming up in the sun. A thick winter coat and a fat layer help insulate the macaques, but out of the hot springs, huddling is the main way to keep warm in extreme cold.*

Jigokudani owes its name to its many hot springs and has long been a favoured haunt for both macaques and humans. Back in 1964, a monkey park was established here, and a pool was subsequently built purely for the macaques, to stop them from invading nearby hot tubs and human spas. Over time, it has become a highly prized asset for rest and recreation, with mainly higher-ranking individuals monopolizing it.

The macaques have a body temperature of about 38°C (100.4°F) and seem to favour a tub temperature of about 41°C (106°F). In these comforting waters, high-ranking youngsters swim and play or suckle from their mothers, and adults and young spend a lot of time grooming and delousing one another.

When they emerge from the hot spring, they don't lose heat as fast as humans would, because they have very few sweat glands and excellent insulation. But their choice of winter refuge is most definitely a very human one, and for Japanese macaques it can make the difference between life and death.

Left Resting at dawn before beginning to feed again on buds, bark and what meagre pickings they can find in the winter woodland. This is as far north as it is possible for a non-human primate to live.

Next page, left Young macaques taking a rest from playing. Winter games can include snowballs and snow-rolling.

Next page, right Taking advantage of a morning sun spot, curling up toes and fingers to avoid frostbite.

Bands, troops and harems

Above *A female attempting to take back her baby from a male who is playing with it. She may need to call on her harem leader's help to get it back. It's in her interest to be with a strong leader who can provide her with protection from other males as well as predators.*

Below *The leader grooming one of his females, strengthening the bonds between them.*

Opposite *A hamadryas group warming up after a night on the rocks, along with more than a hundred other baboons. They rely on safety in numbers to protect them from leopards and hyenas.*

One of the most complex, multilevel social systems of all non-human primates is that of the hamadryas baboon. The northernmost of Africa's baboon species, the hamadryas is found in semi-desert regions in the Horn of Africa, a corner of the Arabian peninsula in Yemen and southwestern Saudi Arabia. Its social structure is very different from those of the other five baboon species, probably moulded by the harsh environment and the need to defend feeding areas, but also by the need to gain protection from predators, which has led to huge congregations, especially at sleeping areas.

The basic grouping is a harem, comprising one leader male and a collection of females and their young, though there may also be one or more 'follower' males. Clans of two or more one-male units group together in the day to form bands, which also contain solitary adult males and immature males. When there is contact between the bands, it's often aggressive, but several bands of up to 400 individuals may join up in the evening to form troops of nearly a thousand individuals, congregating on cliff-faces or rocky sleeping areas.

Some of the most comprehensive studies of hamadryas baboons have been carried out around Filoha, an area of hot springs in the north of Ethiopia's Awash National Park. Here the baboon bands move within a home range of at least 30 square kilometres (12 square miles) of semi-arid acacia thorn-scrub. As they leave the sleeping cliffs, the bands will often travel as a troop, sometimes for more than a kilometre, before taking their preferred feeding routes. The outer layers of doum-palm fruits, acacia leaves, flowers, seeds and pods, and grass seeds, blades and flowers are all part of their staple diet, but when the opportunity arises, they will also chase down Abyssinian hares or snatch swarming locusts.

As they travel, a leader male governs his harem aggressively, herding females who wander too far away

or socialize with rival outsiders. He uses visual threats but will also bite violently, often on the neck or head. The females regularly groom their leader and may even fight over grooming access, especially in times of social turmoil or danger, and those in reproductive condition (oestrus) spend more time close to the male. So do females with infants under two months old, seeking protection from their leader. These bonds may last for years before the male eventually 'steps down' in favour of a younger follower from the unit. Sometimes this handover seems almost voluntary, to an individual the leader has got on well with or is perhaps a relative.

Leader males are generally respectful of other one-male units, communicating with ritualized facial gestures. But every now and then fierce unit takeovers occur, resulting in serious injury to the usurped leader and

sometimes even the death of his infants. Females whose babies are killed will usually become sexually receptive within two weeks of the takeover and then mate with their new leader. Non-leader males may grab infants from their mothers, usually to play with them. Mothers who have lost their infants in this way are generally unable to retrieve them unless aided by the leader male. In times like this, and when danger threatens from rival baboons or predators, the leader can be a reassuring protector and ultimately a very important factor in a female's reproductive success.

In fact, it's in a female hamadryas baboon's reproductive interests to affiliate with a leader male able to provide protection for her and her infants, both from predators and from other baboons competing for food in the harsh desert environment.

Nut-cracking red-faced swingers

The Yavari river basin in northeast Peru has a diversity of monkeys, with 13 recorded species, from miniature pygmy marmosets to lanky black spider monkeys. The strangest looking of them is the Peruvian red uakari – one of four races of red uakaris found in Amazonia.

Red uakaris (pronounced 'wakaris') live in some of the wettest and most inaccessible parts of the Amazon rainforest, among the aguajal-palm swamps and seasonally flooded varzea forests. They often unite in aggregations of up to 200 individuals in a complex social system, with smaller foraging parties constantly departing and reuniting, keeping in contact through raucous *hic-hic-hic* chit-chattering that sounds like laughter. The core of a uakari group comprises females and young, led by a male and often followed by groups of juvenile and subadult bachelors. There is frequent conflict between the bachelors and the ruling male, who tends to form alliances, flaunting his authority by swaying on branches from above or dangling by his ankles with fur bristling, almost doubling his size.

Red uakaris specialize in eating fat-rich seeds. From May to September they have a penchant for the fatty, yellow pulp of ripe aguaje-palm fruits, but when these are exhausted, they move to the forest in search of large-seeded fruits in the tree canopy. Between January and April, when the varzea floods, the uakaris search out seeds of fruits such as those of *Eschweilera* trees before they ripen and fall into the water. The problem is that most of these fruits protect their ripening seeds with extremely hard shells. The uakaris' solution is big canines and enlarged temporal muscles – especially prominent in males – which give them incredibly strong jaws that enable them to crack open the toughest seed casings. Their long incisors then become useful tools for scraping out the seeds.

Most other monkeys can't crack open these unripe fruits, giving the uakaris an advantage. But such fruits tend to be more evenly spread through the forest than the ripe-fruit hotspots where other monkeys tend to

Above *A mother being groomed by her infants in the forest canopy. Like the males, she also has strong jaws and incisors to crack open tough nuts and extract the contents.*

Opposite *A male red uakari in the swamp forest extracting pulp from the fruits of the aguaje palm. In the Yavari river basin, they rely heavily on this food.*

gather, and so uakari groups often travel long distances each day in search of these high-quality seeds.

The lower reaches of their chosen forest home are not only often flooded but are also full of spines and roamed by anacondas, and so red uakaris prefer to spend most of their time in the middle or upper reaches of the forest. Though it may be safer up there, it makes getting around challenging, especially with a short stumpy tail that offers little help with balance. The uakaris rely on skilful leaping, repeatedly rocking branches back and forth to generate momentum before catapulting themselves as far as 6m (20ft) between tree crowns.

It's thought that the maintenance of a bright red face signals to potential mates good health, fitness and resistance to disease – particularly important given that uakaris live in swampy habitats where malarial mosquitoes and other blood parasites breed.

Red uakaris will often associate with other monkeys including capuchins, squirrel monkeys and, especially, woolly monkeys, probably to combine watchful eyes for predators such as harpy eagles, ocelots and tayras. That the different monkeys can all coexist in the varzea forest is because each species feeds at different levels and on different foods.

A hard nut to crack

Growing up is always hard. For young bearded capuchins living in the forests of central Brazil, it's particularly hard. They have to learn to perform a complicated sequence of tasks before they can begin to profit from their most reliable source of food.

Below the cathedral-like sandstone cliffs of the Boa Vista valley, the woodland floor is littered with table-sized rocks with relatively smooth tops. The tops are not flat but marked by small, shallow depressions, like hands gently cupped. These huge blocks of sandstone are anvils. And sitting on the anvils are large, well-polished stones of an altogether different geology. These are the hammer stones, brought to the anvils by the bearded capuchins.

The cliffs, situated in this green, wooded valley, provide the capuchins with a secure retreat at night. Though there is food there, the richest source – an abundance of palm nuts – is several kilometres from the cliffs. Palm-nut kernels offer a valuable source of nutrition, but extracting the kernels takes an extraordinary amount of planning, coordination and, ultimately, effort.

The first step is to harvest the palm nuts. Though they grow conveniently near the ground, they ripen slowly. But adult bearded capuchins know what to do. They tap the nuts with their fingers to test for ripeness, and the ripest nuts are wrested from the plant. They then retreat up a tree – a safer place for a monkey – and with their teeth gradually remove the nuts' fibrous husks. Then, surprisingly, the nuts are discarded. It's possible that the capuchins know that the nuts aren't quite ripe yet and need to be left exposed to the sun for many days before they're ready to eat.

Scattered over the ground are nuts from previous visits to the palms. The bearded capuchins now pick their way through the older nuts, tapping them together or on the ground to check their ripeness. Gathering the favoured nuts in one arm, a capuchin makes the long journey back to the anvils just below the cliffs. It would seem that, by peeling and then discarding new nuts each time the harvesting site is visited, the production line is kept up and each trip is profitable.

The next task at the anvils is to hammer open the nuts. The story of the hammer stones is still little understood. They don't originate from the relatively soft sandstone of the anvils. This is just as well – otherwise they would gradually break up with continued use. They have probably fallen down the ravines and gullies of the valley from conglomerate layers and then been retrieved by the capuchins. Some are made from metamorphosed sandstone, and the hardest ones from quartzite. They are large and usually weigh between a third and a half of an adult bearded capuchin's body.

Left *Hammer, anvil, nut and cracker. The choice of hammer – here probably at least a third of the weight of the bearded capuchin – is crucial, as is the fit of the nut into the depression on the surface of the rock anvil. The tail is useful as a stabilizer.*

Opposite *Testing the ripeness of nuts and gathering them up to take to the anvil site.*

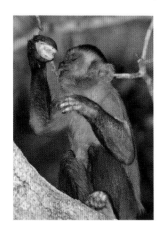

Above *The prize for nut-cracking: a drink of palm-nut juice followed by a feast of kernel.*

Opposite *Nut-cracking school. An adolescent at the anvil, learning from an experienced adult how best to crack nuts.*

A capuchin will place a nut in an anvil depression. Get the wrong hole, and the nut shoots off into the bushes. Get the right hole, and the prize is the kernel. But even an experienced adult rarely cracks such a hard nut in one go. It is hit once, inspected, turned, placed again, hit again, and so it goes on, with deft adjustments of the position of the nut and the grip of the hammer. The amount of energy in each hit varies enormously, also depending on the size and strength of the monkey. Sometimes merely lifting the stone using arms and shoulders is enough. Other nuts require more effort, and a capuchin will stand right up, lifting the rock high above its head before bringing it crashing down on the nut. Given the weight of the hammer stones, this requires an extraordinary amount of strength.

The sound of the nuts being smashed can be heard some distance away – and predators are no doubt alert to the sound. It's therefore not surprising to find that the anvil stones are always under trees. That way, if danger threatens – and the capuchins are always on the lookout for it – there is an easy escape route.

Clearly the young bearded capuchins have much to learn. They follow the adults closely, watching every move, from the selection of a nut, to the stripping away of the husk, to the choice of a sun-ripened nut and finally to the use of the hammer. Then they try it themselves – for months, or even longer. Their attempts are often comical. Youngsters sometimes bang nuts together, much as a young child would bang toy bricks together, or they carefully place a nut on an anvil and then vainly try to break it with another nut. Gradually, though, they learn from the adults (who may give them partially opened nuts to work on), refine their techniques and start using the tools correctly.

This remarkable technique of tool-use must have been passed down from generation to generation over hundreds of years. And like all primate tool-use, it gives us another piece of the jigsaw in our understanding of our own evolution.

Chimp culture, dexterity and devices

The most accomplished of all the animal tool-users, apart from humans, are chimpanzees. Each community has its tool-use culture, fashioning different tools for different tasks. In Bossou in southeastern Guinea, 24 different tool-use behaviours have so far been recorded, serving purposes from pounding and probing to extracting and displaying. And two behaviours – pestle-pounding and algae-scooping – have been recorded only from this chimpanzee community.

The local Manon people revere the chimpanzees of Bossou as reincarnations of their ancestors, inhabiting the sacred forest of Mont Gban, which overlooks the village. The 13 chimpanzees of the current community forage mostly within 6 square kilometres (2.3 square miles) of mainly secondary forest, bordered by a patchwork of cultivated and abandoned fields and riverine and scrub forests. They feed on as many as 200 plant species – some 30 per cent of the available species – with fruits as their mainstay, though they also eat plenty of leaves, pith, seeds, flowers, roots, gum and bark. They supplement this diet with insects, birds' eggs, honey and the occasional mammal. When natural food is scarce, they plunder the orchards and fields for oranges, mangoes, cassava, corn, papayas and bananas, and feed on oil palms, often sharing the bounty – a rare practice among other chimpanzee communities.

Such a variety of food sources has given rise to a wide repertoire of tool-use, especially when wild fruits are scarce. The most sophisticated and best known, which has been observed since studies of the Bossou chimpanzees began in the mid-1970s, is the use of a stone hammer and anvil to crack open oil-palm nuts, whose kernels are rich in energy, protein, calcium, phosphorus, fatty acids and vitamin A. This requires

Above and left A chimpanzee practising the best-known of the Bossou repertoire of tool-use: a hammer and anvil used to crack open oil-palm nuts. Such nut-cracking requires great dexterity and hand-eye coordination.

It's not just the nuts of the oil palm that the Bossou chimpanzees rely on. They also feed on its stalks and flowers and the palm heart. Extracting a heart requires tool use – seen nowhere else. A chimpanzee will climb into the crown of the oil palm, spread out the leaves and, with great force, tug out the central leaves to get to the growing point. Then it may modify one of the leaf stems into a pestle to pound the crown and excavate the softened, juicy palm heart, rich in vitamin B. This happens in the wet season, as does another tool-use practice unique to Bossou – algae-scooping – when a chimpanzee will select a stalk, remove the leaves and use it as a 'fishing rod' to lift algae from ponds.

A wand may also be used to dip for ants. A flexible stem or stick is held between the index and middle fingers and delicately brushed from side to side to encourage the ants to attack. The wand with its clinging ants will then be pulled through the mouth. Or the chimp will swiftly pull the wand through its hand and scoop the ants into its mouth. A stick may also be fashioned to extract bees from dead wood. Other uses for plant materials include folding leaves into 'cups' for drinking water. And when a siesta beckons and the ground is damp, chimpanzees may arrange leaves into comfortable mats. Such inventions are passed from one generation to another as part of the Bossou culture. And as new challenges are presented, new uses will be devised, just as among human societies.

Above **Trying out a stick – useful as a tool or weapon.**

Below **A four-year-old chimp learning from her mother how to use a hammer and anvil. The critical period for learning any such skill seems to be before seven years old.**

excellent dexterity and hand-eye coordination. It involves careful positioning of three movable objects, the nut, the hammer and the anvil, and sometimes even placing a stabilizing stone under the anvil. A proficient adult may crack open three or four nuts a minute. Five- to eleven-year-olds spend a lot of time practising. Not all take part, possibly because there is a critical period for learning – young who don't start to crack nuts before the age of seven don't seem to acquire the skill.

Above *Making use of an adapted liana swing.*

Left *Using a palm-frond stem as a pestle to pound and soften the palm heart. The stem is hammered down with force and then used to dig out the heart. Such enterprising uses of forest resources are passed down the generations as part of the culture of this group of chimpanzees.*

Far left *Fishing for algae with a specially selected rod.*

Index

Credits

Production Team
Bridget Appleby
Rupert Barrington
Simon Blakeney
Jesse Bliss
Adam Chapman
Tom Clarke
Bobbie Fletcher
Mike Gunton
Justine Hatcher
Martha Holmes
Chadden Hunter
Tara Knowles
Neil Lucas
Stephen Lyle
Vivienne Mackay-Hope
Patrick Morris
Emma Napper
Ted Oakes
Lisanne O'Keefe
Victoria Ribeck
Kate Roberts
Elly Salisbury
Adam Scott
Lisa Sibbald
Jonathan Smith
Ian Syder
Rosie Thomas
Nikki Waldron
Barbara Wetheridge
Robert Wilcox
Paul Williams
Emily Winks

Camera Team
John Aitchison
James Aldred
Doug Allan
Doug Anderson
David Baillie
Ralph Bower
Jim Brandenburg
Barrie Britton
John Brown
Keith Brust
Gordon Buchanan
John Chambers
Chris Chanda
Rod Clarke
Dany Cleyet-Marrel
Martyn Colbeck
Bob Cranston
Bruce Davidson
Stephen de Vere
Rudi Diesel
Jason Ellson
Justine Evans
Tom Fitz
Kevin Flay
Richard Ganniclifft
Ted Giffords
Ian Goldsbrough
Nick Guy

Charlie Hamilton James
Mike Holding
Adam Huddlestone
Richard Jones
Simon King
Richard Kirby
Peter Kragh
Mark Lamble
Yves Lefevre
Alastair MacEwen
Dave MacKay
Jamie McPherson
Justin Maguire
Michael Male
Dave Manton
Richard Matthews
Charles Maxwell
Hugh Maynard
Hugh Miller
Shane Moore
Roger Munns
Peter Nearhos
Mark Payne-Gill
Andrew Penniket
Steve Phillipps
Mike Pitts
Steven Romano
Rick Rosenthal
Peter Scoones
Tim Shepherd
Andy Shillabeer
Warwick Sloss
Mark Smith
Sinclair Stammers
Ian Thomas
Gavin Thurston
Simon Werry
Peter West
David Wright
Norbert Wu
Mark Yates
Kazutaka Yokoyama

Field Assistants
Graham Abbott
Dave Boguski
Anthony Bramley
Paul Brehem
Chris Browne
Jim and Stephanie Carpenter
Wirong Chanthorn
Al Coldrick
Bryan Curran
Alicia Decina
Ben Dilley
Georgette Douwma
Rob Dover
Stephen Dunleavy
Will Engleby
Ethiopian Wolf Conservation
  Programme
Edmund & Kim Farmer
Tim and Pam Fogg

Richard & Carol Foster
Camila Galheigo Coelho
Angel Garcia-Rojo
Mulualem Gelaye
Berhanu Geremew
Daniel Gomez
Lance Goodwin
Vinita Gowda
Annie & Ian Gray
Rob Harvey
Simyra Hlebechuk
Bruce Inglangasak
Jason Isley
David Jones
David Karanja
John Keeling
Komodo National Park
  rangers
Duncan Mackay
Julio Madriz
Gil Malamud
Marie Louise
Umaporn Matmeen
Joseph Mfune
Andrew Miners
Robert Morrison
Sammy Munene
Cameron Newall
Lasse Østervold
Tuomas Palojärvi
Patrick Plantard
Jerome Poncet
Gilbert Rakotoarisoa
Bertrand Razafimahatratra
David Rootes
Norbert Rottcher
Alan Rowley
Jo Ruxton
Hiroo Saso
Jenny Sharman
Digpal Singh
Maguerite Smits Van Oyen
Peter Snyman
Lisa Solberg
Matthew Swarbrick
Mark Thurlow
William Trim
Richard Uren
Chris White
Emilio White
Like Wijaya
Ben Winger
Chanpen Wongsripheuk
Stephen Yiasoi

Scientific Consultants
Richard Bodmer
Mark Bowler
Warren Brockelman
Lincoln Brower
Kevin Campbell
Pompilio Campos Chinchilla
Victoria Cartledge

Ken Catania
James Cosgrove
Sam Cotton
Marta de Ponte Machado
Louis du Preez
Marion East
Laura Engleby
Libby Eyre
Chris Fallows
Lisa Filippi
Rachel Graham
Randy Griebel
Michael Guinea
Karina Hall
Dagmar Hilfert-Rüppell
Ben Hirsch
Kimberely Hockings
Elizabeth Hofer
Tatyana Humle
Atsushi Ishimatsu
Shiguyuki Izumiyama
Jack James
Kevin Kalasz
Jonathan Kingdon
John Kress
Darryl Kuamo'o
Eileen Larney
Matthew Lewis
Stanislav Lhota
John Louth
Sue Mansfield
Robert Mason
Tetsuro Matsuzawa
Iwein Mauro
Clayton May
Bruce Means
David Mitchell
Santos Montenegro
Brian Morland
Robert Nishimoto
Lance Nishiura
Shintaro Nomakuchi
Harri Norberg
Justin O'Riain
Don Owings
Gail Patricelli
Mat Pines
Pang Quong
Gilbert Rakotoarisoa
Galen Rathbun
Barry Rice
Heidi Richter
Flavio Roces
Lynn Rogers
Dagmar and Georg Rüxppell
Myron Shekelle
Wade Sherbrooke
Michael Sheriff
Leigh Simmons
Stephen Simpson
Gabriel Skuk
Jennifer Small
Dee Snijman

Takayo Soma
Emma Stokes
Larissa Swedell
Ethen Temeles
Angelique Todd
Sumio Tojo
Ray Townsend
Alfredo Ugarte-Peña
Sri Suci Utami Atmoko
Elisabetta Visalberghi
Caroline Yetman

Post-production
Bridget Blythe
Linda Castillo
Janne Harrowing
Ruth Peacey
Sarah Wade
Georgina Way

Music
George Fenton

Film Editors
Nigel Buck
Andrew Chastney
Martin Elsbury
Darren Flaxstone
Andy Mort
Andy Netley
Jo Payne
Dave Pearce

Sound Editors
Kate Hopkins
Tim Owens

Dubbing Mixers
Chris Domaille
Graham Wild

Colourist
Luke Rainey

Graphics
Burrell Durrant Hifle (BDH)
Mick Connaire

Picture Online
Tim Bolt
Fred Tay

Discovery Channel
John Cavanagh
Susan Winslow

The Open University
Sally Ashwell
Catherine McCarthy
David Robinson
Janet Sumner

Acknowledgements

Commissioning Editors Shirley Patton and Muna Reyal
Project Editor Rosamund Kidman Cox
Designer Bobby Birchall, Bobby&Co Design
Picture Researcher Laura Barwick
Production Manager David Brimble

Colour origination by XY Digital Ltd

Printed and bound in the United Kingdom by Butler Tanner and Dennis Ltd

The making of the television series *Life* and the writing of this book are inextricably linked. The authors are, by and large, the programme producers, and they would, without reservation, like to thank everyone who was involved with this monumental project. It has taken nearly five years to bring *Life* to life, and throughout that time, countless people have given their support, enthusiasm, time and energy freely. There are too many people to whom we owe a debt of gratitude to mention here, but to all, we are eternally grateful. Some of those directly involved in the filming and making of the series are listed.

For three years, teams were filming for *Life* around the world. We filmed on every continent and made more than 150 filming trips. Each trip involved consultation with a scientist or expert, logistical planning by the production team, weeks in the field for the camera teams, sometimes in appalling conditions, sometimes in relative comfort.

Many of *Life*'s successes were due to the generosity of people on location. Some have given up huge amounts of their time, others willingly let us share their resources. Filming Komodo dragons hunting and fighting would not have been possible without the endless support of Marcus Matthew-Sawyer and his team at Pt. Putri Naga Komodo. Al Coldrick ensured that we captured male humpback whales fighting over a female in Tonga. Lincoln Brower shared his decades of research to ensure the monarch butterfly shoots were successful. In Antarctica, the Royal Navy and the crew of the HMS *Endurance* generously gave us helicopter time so that we could film killer whales hunting among the ice floes from the air. The National Science Foundation supported our work on the Antarctic continent, giving us access to a world beneath the ice. The inimitable Jerome Poncet led an expedition that saw land-based and diving teams strung out along the peninsula. And so it is around the world, the list is almost endless. We are extremely grateful to everyone involved.

And, of course, there are those responsible for the book: Shirley Patton was kind enough to commission it; Muna Reyal has overseen the project; Laura Barwick has found many breathtaking photographs; Bobby Birchall designed the book; and Roz Kidman Cox has had the patience and passion to ensure this book is as beautiful and informative as it should be. We thank them all.

Picture credits